Messrs. Methuen's New Books

Spring, 1923

GENERAL LITERATURE

ON: A Volume of Essays. By HILAIRE BELLOC. F'cap. 8vo. 6s. net.

THESE LIBERTIES. By "EVOE" (E. V. KNOX). F'cap. 8vo. 4s. 6d. net.

ENGLISH DIARIES. By ARTHUR PONSONBY, M.P. Demy 8vo. 21s. net.

THE SECOND PART OF KING HENRY IV. Edited by R. P. COWL. Demy 8vo. 6s. net. [*The Arden Shakespeare*

MUCH ADO ABOUT NOTHING. Edited by GRACE TRENERY. Demy 8vo. 6s. net. [*The Arden Shakespeare*

SHELLEY: The Man and the Poet. By A. CLUTTON BROCK. A new Edition. F'cap. 8vo. 7s. 6d. net.

A HISTORY OF FRENCH LITERATURE. By KATHLEEN T. BUTLER. Two volumes. Each Crown 8vo. 10s. 6d. net.

THE CHILDREN OF THE SUN: An Enquiry into the Early History of Civilization. By W. J. PERRY, M.A. With Maps. Demy 8vo. 21s. net.

ROMAN POETRY. By E. E. SIKES, M.A. Crown 8vo. 8s. 6d. net.

A HISTORY OF GERMAN VERSIFICATION: Ten Centuries of Metrical Evolution. By HENRY GIBSON ATKINS, D.Lit., M.A. Crown 8vo. 10s. 6d. net.

THE BEST OF HAZLITT. Selected and Edited by P. P. HOWE. F'cap 8vo. 3s. 6d. net.

SAILOR-TOWN DAYS. By C. FOX SMITH. Illustrated. Crown 8vo. 6s. net.

YULETIDE REVELS IN MERRIE ENGLAND. Arranged for Acting by M. A. LEWIS and E. M. VERINI. Illustrated. F'cap. 8vo. 2s. 6d. net.

GREAT AND SMALL THINGS. By SIR RAY LANKESTER, K.C.B., F.R.S. Illustrated. Crown 8vo. 7s. 6d. net.

NEW PLAYS FROM OLD TALES. By HARRIET S. WRIGHT. Illustrated. Wide Crown 8vo. 6s. net.

THE MYSTERY OF THE HIVE. By EUGENE EVRARD. Translated by BERNARD MIALL. Crown 8vo. 7s. 6d. net.

THE PRINCIPLES OF AGRICULTURE. By Major J. R. AINSWORTH-DAVIS, M.A. Illustrated. Crown 8vo. 6s. net.

TOWN PLANNING AND THE CITY OF THE FUTURE. By S. D. ADSHEAD, Professor of Town Planning in the University of London. Illustrated. Demy 8vo. 10s. 6d.

SINGLE BLESSEDNESS, AND OTHER OBSERVATIONS. By GEORGE ADE. Crown 8vo. 6s. net.

THE SHAKESPEARE GARDEN. By ESTHER SINGLETON. Illustrated. Large Crown 8vo.

ILLUSTRATED BOOKS

REMBRANDT'S PAINTINGS: With a Commentary and an Essay on his Life and Work. By D. S. MELDRUM. With 544 Illustrations. Wide Royal 8vo. £3 3s. net. [*Classics of Art*

THE NEW PHYSICS

THE NEW PHYSICS

LECTURES FOR LAYMEN AND OTHERS

BY

ARTHUR HAAS, PH.D.

PROFESSOR OF PHYSICS IN THE UNIVERSITY OF LEIPZIG

AUTHORISED TRANSLATION
BY
ROBERT W. LAWSON, D.Sc., F.INST.P.

WITH SEVEN DIAGRAMS

METHUEN & CO. LTD.
36 ESSEX STREET W.C.
LONDON

This Translation First Published in 1923

PRINTED IN GREAT BRITAIN

PREFACE TO THE ENGLISH EDITION

IN the lectures contained in the following pages, it has been my endeavour to describe the picture of nature revealed by modern physics in the most intelligible manner possible, and without the use of any mathematical formulæ. I have aimed at including no more than is necessary to obtain a clear and general view of the subject, and I have intentionally maintained silence on all other aspects, both as regards facts and names. Short footnotes have been added, however, on points which have an immediate connection with the subject-matter of the lectures.

As compared with the original German edition, which appeared in the summer of 1920, the English edition has been extended by the inclusion of a new lecture on the Theory of the Chemical Elements, thus increasing the number of lectures from five to six. Moreover, the lecture on the Theory of Relativity has been materially extended, inasmuch as Einstein's Theory of Gravitation has been dealt with much more fully, and its cosmological consequences have also been discussed. The advances that have been made in physics in the two and a half years since the appearance of the original German edition have also received attention.

In conclusion, it would be a source of real pleasure to me, were the translation of my little book to render a service to the people of a nation which has produced such men as Newton, Faraday, and Maxwell, to say nothing of those who are still living, such as Sir J. J. Thomson and Sir Ernest Rutherford. Without the great genius of these men physics would never have soared to such prodigious heights.

ARTHUR HAAS.

VIENNA, *February*, 1923.

TRANSLATOR'S NOTE

THE rapid advances made in physics during recent years have made it impossible for the student and the layman to keep pace with modern developments, and most of the books that have been written on the results of contemporary research have been intended more particularly for the specialist. Professor Haas is an eminent authority on and exponent of modern physics, and in "The New Physics" he has supplied a long-felt want by presenting a connected and illuminating account of the present day structure of physics, in such a form that it should prove of value to, and within the reach of the average scientific reader.

In the translation I have endeavoured to adhere to the original as closely as possible. I have benefited from suggestions made by my friends, Professor S. R. Milner, F.R.S., and Dr. W. E. Curtis, who, in addition to Professor Haas, kindly read over the manuscript. I take this opportunity

of expressing to them my thanks, as well as to the publishers for their co-operation and careful attention whilst the book was in the press.

A word of explanation is perhaps necessary for the retention of the terms "positive electron" and "negative electron" in the text, instead of the words "proton" and "electron" which are now usually adopted in modern physics in this country. As this book is intended primarily for the layman, it was felt that the former expressions would probably be more intelligible, in particular since the introduction by Sir Ernest Rutherford of the word "proton" for the nucleus of the hydrogen atom is of comparatively recent date, and has not yet found universal acceptance (Sommerfeld still uses "positive electron" and "negative electron" in the most recent edition of his "*Atombau und Spektrallinien*"). However, it is to be desired that uniformity of nomenclature may be attained in the near future, and that the word "proton" will be adopted for the positive elemental "brick" of matter, alongside the negative element—the "electron."

Frequent reference to very large numbers is made in the text, and as the designations used in France and America differ from those used in this and in most other countries, I append an explanatory table. The use of the mathematical notation did not seem quite justified in a semi-popular book, and for that reason the words in the text are written out in full. The French and American equivalents are given alongside in brackets :—

Million .	$1,000,000 = 10^6$	; (*million*).
Milliard .	$1,000,000,000 = 10^9$	; (*billion* or *milliard*).
Billion .	$10^6 \times 10^6 = 10^{12}$	; (*trillion*).
Trillion .	$10^6 \times 10^{12} = 10^{18}$	; (*quintillion*).
Quadrillion	$10^6 \times 10^{18} = 10^{24}$	; (*septillion*).
Quintillion	$10^6 \times 10^{24} = 10^{30}$	;
Sextillion	$10^6 \times 10^{30} = 10^{36}$	;
Septillion	$10^6 \times 10^{36} = 10^{42}$	;
Octillion .	$10^6 \times 10^{42} = 10^{48}$.	

R. W. LAWSON.

TABLE OF CONTENTS

THE NEW PHYSICS

I. THE ELECTROMAGNETIC THEORY OF LIGHT

IN consequence of the magnificent advances made in theoretical physics since the beginning of the twentieth century, our views on natural phenomena have undergone a complete transformation. The foundations of Natural Science have been revolutionized. Deeply rooted conceptions have been shown to be untenable prejudgments and have been condemned. The oldest ideas of natural philosophy have changed their meaning. Previously unsuspected relationships have been laid open to physical research. Our conception of nature has been beautifully extended, and at the same time simplified and unified.

This most recent development of physics, so revolutionary and yet so successful, was aided by two theories which originated in the latter half of the nineteenth century, each of which represented a great advance in an endeavour to obtain uniformity in physics. The first of these theories, due to **Maxwell,** identified the phenomena of light with those of electricity, and thus converted **optics** into a **branch of the science of electricity.** The other theory explained the phenomena of heat by assuming a continuous motion of the smallest particles composing bodies, and thus the **science of heat** became a **branch of mechanics**. At the same time, this latter theory clearly showed the extraordinary fruitfulness of the **atomistic** principle. At the end of the nineteenth century the extension of Maxwell's theory led to the **electron theory**. This theory enabled us to regard electricity as the primordial constituent of all things, and to recognize in small electrical charges the building stones of matter. The **quantum theory**, which constitutes an extension and a development of the atomistic principle, has been pre-eminently successful, especially in the interpretation of spectra, and it has supplied us with valuable information on the internal structure of the **atoms of the chemical elements**. The **theory of relativity** brought about an entirely new conception of the nature of space and time, of motion and of matter, and from a generalization of the theory of relativity we have been led to the modern **theory of gravitation**, which also yields important information on cosmological questions.

The first of these lectures will be devoted to Maxwell's **electromagnetic theory of light**, and their object throughout will be to give a readily intelligible presentation of the modern physical conceptions of natural phenomena.

In **optics**, as in every branch of physics, there are certain **fundamental phenomena**, amongst which the rectilinear propagation of light, its reflection and its refraction have been known since remote antiquity.[1] Numerous fundamental phenomena in optics were

[1] The law of reflection was discovered in ancient times, but the law of refraction was not discovered until the middle of the seventeenth century by Snellius.

discovered in the second half of the seventeenth century. Of these we may mention the diffraction of light, its double refraction in crystals, the remarkable colours of thin films, the scattering of light, and finally, the fact of a perfectly definite velocity of propagation of light. At that time the question of the nature of light occupied the attention in particular of the two leading physicists, **Newton** and **Huygens.** Although they held different views [2] as to the nature of light (details of which cannot be given here), they nevertheless agreed on one important point, namely, that **light** must be a **periodic** process both **as regards time and space.**

We can conceive of such a double periodicity in the following manner. Let us consider a particular position in space, and suppose that a definite series of conditions is repeated periodically at this place. The number which states how often the succession of conditions is repeated in the unit of time is called the **frequency.** [3] Instead of comparing the conditions

[2] During the eighteenth century and the first two decades of the nineteenth century, the undulation and the emission theories vied with each other for precedence. The former regarded light as being mechanical waves, whilst the latter regarded it as substance in the form of very small particles. Newton certainly laid the foundation for the emission theory, although he intentionally avoided taking a stand for any definite hypothesis. But we should be quite unjustified were we to disparage Newton's services to theoretical physics on this account. Newton's important discovery of the periodicity of light is independent of all special conceptions, in fact, from the modern viewpoint, we must give preference to a theory which referred to periodicity in a general sense and not specially to mechanical waves.

[3] A complete period exists between two like conditions or states when not only the conditions themselves are alike, but also the direction (sense) of their variations is the same. We may express this mathematically by saying that in both cases the differential

in one and the same place at different times, we may compare the conditions in different places at the same instant of time. For this purpose we imagine a straight line drawn from some particular position in space, and we compare with each other the conditions that exist at different places along this straight line. This may be done, as it were, by taking an instantaneous photograph. We then have a spatial picture of conditions along this straight line. If, now, the process is of such a kind that the **temporal succession** of the conditions at one and the same place is **the same** as the **spatial succession** along such a straight line at a particular instant of time, we designate (in the broader sense of the word) such a process of double space- and time-periodicity as a **wave**.[4] Along a straight line, or, as we also say, along a **ray** we find the same succession of conditions periodically at definite distances apart, each of which we call a **wave length.** Moreover, the **velocity of propagation** of the wave is simply the product of the frequency and the wave length.

The best known examples of undulatory processes are those which consist of **processes of oscillatory motion**, such as are brought about when one throws a stone into a lake, or when one blows a whistle, and thus sets the air into vibration, giving rise to what our ear recognizes as a **tone**. Until quite recent times, physics endeavoured to **explain** all

coefficient with regard to time of the entity in terms of which the condition is measured must have the same sign.

[4] This definition is not quite exact. It only serves the purpose of making the difficult modern conception of a wave more easily intelligible to the layman. For the more exact definition reference may be made to § 55 of the author's "Introduction to Theoretical Physics" (Constable).

physical phenomena as far as possible **in terms of processes of motion.**[5] We can thus understand how it was that **Huygens** endeavoured to interpret his assumed waves of light as **mechanical** ones.[6] But since light is propagated from the sun to the earth through a manifestly empty space, and on the other hand we should find difficulty in otherwise explaining the immense value of the velocity of light,[7] we could not regard light as the wave motion of any ordinary medium. Huygens had no alternative, then, but to make that hypothetical, mysterious æther the carrier of the light waves, although the æther had been conceived before this by other thinkers and for other reasons.[8] As in solid and liquid bodies, so also in the æther, according to Huygens' idea, it should be possible to have **mechanical waves** founded on **elastic vibrations**, which we experience as light.

Science owes a great and a lasting debt to Newton and Huygens, for they were the first to recognize the **space-time periodicity** and thus (in the wider sense of the word) the **wave nature of light**. The fact

[5] Characteristic of this line of thought is Huygens' dictum, that in real science we can only comprehend the cause of all effects by the mode of thought of mechanics, if we do not wish to renounce for ever all hope of understanding anything whatsoever in physics. The tendency in physics to reduce everything to mechanics was largely due to the philosophy of Descartes which for a long time was so influential.

[6] Huygens thought of light waves as being quite analogous to sound waves in air, and also assumed them to be longitudinal, like sound waves.

[7] The velocity of light is 300,000 kilometres per second. Light thus requires about 8 minutes to traverse the distance between the sun and the earth.

[8] The ether played a great part even in the natural philosophy of Descartes.

that optical waves in particular were regarded as mechanical ones is due historically to the tendency in physics at that time of reducing everything to mechanics, but this was by no means necessary. Newton's and Huygens' idea of the spatial and temporal periodicity of light would have completely sufficed for the great advances which were made in theoretical optics up to the time of Maxwell. This would have been none the less true even if the solution of the question as to the true nature of these doubly periodic processes had been deferred, and scientists had not prematurely decided in favour of the special assumption of the mechanico-elastic nature of optical processes.

Between 1808 and 1820 a new group of optical phenomena was discovered, **the phenomena of polarization**. The discovery of these phenomena made it clear that the principle of the space-time periodicity of light was capable of giving an extraordinarily simple explanation even for quite complicated optical phenomena. The phenomena of polarization found a very simple explanation on the optical wave hypothesis, but not until this had received an important and fundamental extension by virtue of the particular assumption of the **transverse nature of light waves**.

In order to make this idea clear in its modern significance, it is necessary to discuss first another idea which is of the greatest importance for the new physics. This is the idea of a **vector magnitude**. When we say that the temperature is 15° at a particular place, then, as soon as we have been given a definite **scale** for the measurement of the temperature, the temperature at the place is completely determined by this statement of a single number.

But there are also magnitudes in physics which are by no means completely determinable from the statement of a single number. There are magnitudes that are not known to us until we know their **direction**, as well as their amount. As instances of such magnitudes we may mention velocity, acceleration, and force. They are characterized by the fact that they can be **represented symbolically by a directed interval** of definite length and of definite direction. Since directed intervals are known in geometry as **vectors**, we designate as **vector magnitudes**[9] such physical magnitudes as can be represented symbolically by a directed interval or line.

From the experimental observations of the above-mentioned optical phenomena of polarization the following remarkable conclusion necessarily resulted: The optical condition at a position in space must be capable of representation by a vector magnitude which varies periodically, and which, moreover, is always **perpendicular** to the direction of propagation of the light.[10] It is for this reason that we speak of the transverse nature of light waves.

The theory of light based on this conception, and developed mainly by **Fresnel** in the second and third decade of the nineteenth century was pre-eminently

[9] Contrasted with this, magnitudes which, for a given scale, are determined simply by the statement of a single number, are called scalars. Analytical mechanics is founded on the fact that a vector can also be determined by the statement of three scalars, instead of by the statement of its magnitude and direction. These scalars represent the components of the vector along the three axes of a spatial co-ordinate system.

[10] The transverse nature of light follows from the fact that the ordinary and extraordinary rays into which a ray of light can be split up by double refraction in a crystal cannot under any circumstances be made to interfere, either completely or partially.

successful. But nevertheless it had its weak side, which gradually became a source of great embarrassment. Like Huygens, Fresnel regarded light waves as **mechanical waves.** (According to Fresnel's idea, the periodically varying vector magnitude would be simply the directed interval drawn from the so-called position of rest of an oscillating æther-particle to its position at a particular moment in question.) According to the ideas which prevailed throughout a large part of the nineteenth century, light waves would be determined by the **elasticity of the aether**. At this point insurmountable difficulties arose, because the idea of purely transverse waves stood in direct contradiction to the results of the theory of elasticity, and to the properties which had to be ascribed for other reasons to the hypothetical æther.[11] Although Fresnel's elastic theory of light had proved very useful in the explanation of complicated optical phenomena, its foundations were nevertheless completely untenable and contradictory. The assumption of the wave nature and of the transverse nature of light was correct, but on the other hand the special concept of the mechanico-elastic nature of light waves was erroneous.

It was the English physicist **Maxwell** who first recognized this, and who at the same time also first grasped the true nature of light. Maxwell's

[11] According to the elasticity theory, only longitudinal waves can occur in the æther, if it be a liquid. Transverse waves in addition to longitudinal ones are only possible when we consider the æther to be solid. If we desire entirely to exclude longitudinal waves, we have no other option than to regard the æther as also being incompressible. But it is difficult to make this idea consistent with the necessary assumption that the æther offers no resistance to the motion of the heavenly bodies.

work consisted in the assimilation of optics into the theory of electricity.

Until towards the end of the eighteenth century, the **science of electricity** had been solely the science of **frictional electricity.** In the last few years of the eighteenth century **electric currents** were discovered. In the year 1820 **Oersted** discovered the fact of so-called **electromagnetism,** that an electric current produces in its neighbourhood a magnetic field. Eleven years later, in 1831, **Faraday** succeeded in making the extremely important discovery of **induced currents.** These induced currents arise in a closed conductor situated in a magnetic field when the magnetic field **varies,** and no matter whether the magnetic field is due to an actual magnet or is produced by another electric current.

The exact theory of electrical and magnetic processes is founded on the laws that describe the phenomena of electromagnetism and of induced currents. In the year 1873 Maxwell supplemented this theory by means of a new hypothesis of extreme importance. He carried ideas of Faraday a step farther and assumed that not only in wires, but also in empty space and in electrical insulators electric currents of a kind must exist, and these he designated (for reasons which need no further discussion here) **displacement currents.**[12] Maxwell ascribed to these displacement currents just **the same properties** as are observed in the case of closed **currents in conducting wires.** Thus displacement currents ought to produce a magnetic field, they ought to be called

[12] The reader will find details of the difficult conception of the displacement current in the author's "Introduction to Theoretical Physics," § 47, a translation of which is in preparation (Constable).

forth and influenced by induction, and they ought themselves to be capable of inducing currents.

The introduction of this hypothesis into the theory of electricity and of magnetism led, in a purely **deductive** manner, to the most wonderful conclusions. By the application of mathematical methods, the **theoretical possibility of electromagnetic waves** resulted from the hypothesis of displacement currents, and, as opposed to elastic waves, these electromagnetic waves were necessarily **transverse** ones. How have we to conceive of such electromagnetic waves? In order to answer this question, our purpose will be best served if we derive a concept which is very important in theoretical physics, to wit, that of electric field strength.

If an electrically charged body be situated somewhere in space, it produces in virtue of its charge a so-called **electric field**, which manifests itself in that it exerts an attracting or a repelling force on a so-called test charge brought into its neighbourhood. According to the fundamental law of electrostatics (Coulomb's law), this force is proportional to the magnitude of the test charge, and thus the quotient obtained by dividing the magnitude of the force by that of the test charge is independent of the test charge itself. This quotient is called the **electric field strength** at the place under consideration. The field strength is a vector magnitude in the sense previously mentioned; its direction is identical with the direction of the force which is exerted on a positively electrified test charge placed at the point considered.[13]

If the electric field strength existing at a position

[13] We must expressly speak of positive electrical charge, because only then is the direction (sense) of the field intensity determined.

in space **periodically varies** in its magnitude or its direction or in both of these, we say that **electrical oscillations** are occurring at that place. We may mention in passing that these are called linear when only the magnitude varies periodically, circular when only the direction alters, and elliptical when both the magnitude and the direction vary. If a definite series of values of the electric field strength (in magnitude and in direction) recurs periodically with regard to time at a position in space and in the manner mentioned above, it follows from Maxwell's theory of displacement currents that, at any arbitrarily chosen instant of time, the same sequence of values of the field strength must appear ranged alongside of each other **periodically in space**, along a straight line drawn from that position in space. After what has already been said, such a process of double space-time-periodicity will have to be called an electric wave. The idea of **magnetic** field strength is entirely analogous to that of **electric** field strength. In like manner, the concepts of **magnetic oscillations** and of **magnetic waves** correspond to those of electric oscillations and of electric waves.

On the basis of the previously known laws of electromagnetism and of induced currents, together with his new hypothesis of displacement currents, Maxwell was able, by purely deductive methods, to show the following properties of the theoretically possible electric and magnetic waves : Every electric wave is always and necessarily accompanied by a magnetic wave of equal velocity, and inversely, a magnetic wave without an electric wave is impossible. The magnetic field strength is always **perpendicular** to the electric field strength, and both are perpendicular to their common direction of propagation. Electromagnetic waves are thus pure **transverse** waves. In contradistinction to elastic waves they thus possess the property which must necessarily be ascribed to light waves in virtue of the phenomena of polarization.

But the most remarkable result of Maxwell's theory is the value which is obtained for the **velocity of propagation of electromagnetic waves**; moreover, this question is most closely connected with the question of the **measure** of an electric **current**. We can define the strength of an electric current in two different ways. First, we may define it in terms of the quantity of electricity which passes across the section of the conductor in unit time, and secondly, in terms of the magnetic force set up by the current. Thus, even before Maxwell, physicists knew of **two different systems of measure** (**units**) for electric current strength and other magnitudes connected with it, a so-called electrostatic system and a so-called electromagnetic system. The units of current strength are not the same in these two systems. Moreover, their **ratio** is not a pure number like the ratio between inch and centimetre, but, as is shown by closer investigation, it represents a **velocity**.

By means of a method with which we need not further concern ourselves here, Wilhelm **Weber** (who is also known as the inventor of electric telegraphy) succeeded in the year 1856 in determining this ratio experimentally,[14] and he found for it a value which

[14] Wilhelm Weber and his co-worker Kohlrausch first measured the charge of a Leyden jar in electrostatic units by means of an electrometer, and then discharged it through a specially constructed ballistic galvanometer by means of which they could evaluate in electromagnetic units the total quantity of electricity that had passed through the galvanometer (as time-integral of the current).

deviated inappreciably from the known value for the **velocity of light.** The difference was so slight that it had to be ascribed to the imperfection of the experimental method.

Maxwell's theoretical investigations led to the result that the velocity of propagation of electromagnetic waves in a substance depends on the one hand on this constant of Weber, and on the other hand on those constants which characterize the electric and magnetic behaviour of the substance involved.[15] **In empty space** the velocity of propagation is equal to the **Weber constant**, which in its turn agrees with the **velocity of light.** This remarkable result led Maxwell to the idea of **identifying light waves** with the **electromagnetic waves** which he had recognized as **theoretically possible.** He was strengthened in this idea by the fact that, although it had been an insoluble riddle to the elasticity theory of light, the transverse nature of light vibrations results with mathematical certainty from the foundations of the Maxwell theory. A further distinct advantage of the electromagnetic theory of light lay in the fact that the phenomena of reflection and of refraction of light could be explained by Maxwell's theory without drawing upon any further assumptions. Contrasted with this, these phenomena had been explicable on Fresnel's elasticity theory of light only with the help of several highly artificial hypotheses specially devised for the purpose.

As a further result of Maxwell's theory, it was

[15] These are the so-called dielectric constant and the magnetic permeability. These two magnitudes express by how many times the electric or magnetic field strengths respectively appear reduced in the substance, as compared with the value they would have under otherwise identical circumstances in empty space.

shown that a simple relation must exist between the constants which characterize the optical behaviour and those which characterize the electrical behaviour of substances.[16] Soon after the publication of Maxwell's theory, **Boltzmann**, by means of measurements on gases, showed that the relation thus derived from the theory is actually fulfilled. But the most convincing proof of the correctness of the electromagnetic theory of light was not forthcoming until fifteen years later, when, in the year 1888, **Hertz** published the results of his famous experiments. From these experiments it was clear that, in point of fact, waves can be produced by purely electrical means, and that they are propagated, as could be shown by means of electrical apparatus, according to exactly the same laws and with the same velocity as light waves. Moreover, like these, they could be reflected, refracted, diffracted, polarized, and they also showed the phenomena of interference.

In order to understand the theoretical importance of Hertz' experiments, we must in the first place devote a little time to the objective significance of an optical concept that is of subjective origin. This is the concept of **colour**. Newton had already recognized that the difference in the sensations produced by colour is objectively determined by a **difference in the period** of the light. He had also already recognized that the relative difference in the known colours cannot be a very large one. For **violet** light,

[16] This, the so-called Maxwell relation, states that the dielectric constant of a medium is equal to the square of the index of refraction. This relation is very well fulfilled in the case of gases, but very badly for alcohol and particularly for water, of which the dielectric constant is 81, whereas its refractive index is only 1·33. Compare note 12 of the third lecture.

which constitutes one end of the visible spectrum, he saw that the period must be about **half as large** as that for **red** light, which constitutes the other end of the spectrum. Conversely, the frequency of violet light would be double that of red light. Just as in acoustics we say that a tone with twice the frequency of another tone lies an octave higher than the latter, so in optics we can say that one colour lies an octave higher than another when its frequency is twice as great. Thus the visible spectrum, extending from the red to the violet, comprises approximately **one octave.**

At the beginning of the nineteenth century, and on the basis of the wave theory of light, **Young** was able to determine the absolute values of the wave lengths and the frequencies for the different colours. For this purpose he used the known measurements on the so-called Newton's rings, which are observed when a low power lens is laid upon a glass plate. For visible light the wave length amounts to between four and eight **ten-thousandths of a millimetre.** As can be seen from the enormous value of the velocity of light, the frequencies corresponding to these wave lengths are exceedingly large, and amount to between 800 and 400 **billions in a second.** The period of visible light is thus related to one second approximately as this interval is related to a period of time of about 10 to 20 millions of years.

At about the same time as Young in this manner first calculated the frequencies of light, the conception of light itself also experienced an important extension in a two-fold sense, as a result of two important experimental discoveries. In the year 1800, the famous astronomer **Herschel** made the remarkable observation that the **heating effect** of the sun's light is not confined to the visible spectrum, but extends beyond the red end of the spectrum, and does not reach its maximum until this has been left behind. Herschel was thus the discoverer of the **infra-red rays.** On the other hand, just about the same time, **Ritter** found that the **chemical** action of light extends beyond the violet end, and that beyond the violet end of the spectrum it is even greater than in the spectrum itself. The **ultra-violet rays** were thus discovered by Ritter.

Since that time the improvement of physical instruments and methods has made it possible to investigate very exactly the infra-red and ultra-violet spectra. The former are recognized by their heating effect, and the latter by their chemical action. For the detection of the heating effect we now possess exceedingly sensitive instruments, which make it possible to detect the heat from a candle flame even at a distance of 100 metres.[17] On the other hand, in the photographic plate we have an extremely sensitive instrument for the detection of the ultra-violet rays.

Whereas the visible spectrum comprises only one octave, it has been possible in the investigation of the infra-red spectrum to penetrate eight octaves beyond the visible spectrum. The longest wavelength hitherto detected in the infra-red is of magnitude somewhat less than half a millimetre. Until the year 1920 it had been possible to advance beyond the violet end of the spectrum through about four octaves, but in 1921 **Millikan** succeeded in extending this range by an additional four octaves.

[17] This instrument is the so-called bolometer. Its action is based on the fact that a platinum wire alters in electrical resistance with the temperature.

According to Maxwell's theory, all these light rays of the visible and the invisible spectrum represent none other than electromagnetic oscillations, and they thus appear to be brought into close connection with the **oscillations** detected by the German physicist **Feddersen** in the **discharge of a Leyden jar,** fourteen years before the introduction of Maxwell's theory. By means of a rapidly rotating mirror, Feddersen drew out the photographic image of the spark discharge of a Leyden jar, and was thus able to show photographically not only the oscillatory nature of the discharge, but he was also able to determine the number of oscillations per second. For this number he found from his experiments values ranging from about 50,000 to 1,000,000 oscillations in a second. If we assume in accordance with Maxwell's theory that these oscillations are propagated with the velocity of light, we find that they correspond to wave lengths of the order of magnitude of kilometres, and of course we cannot make direct experimental observations with waves of such length.

By means of different artifices, **Hertz** was enabled to increase the frequency of such electric discharge oscillations quite appreciably. He succeeded in reducing the wave length to a few metres, and with waves of this length Hertz was able to experiment conveniently and to show, as already mentioned, that they have just the same properties and follow the same laws as the waves of visible light. Since the experiments of Hertz, which, as is well known, constitute the foundation of wireless telegraphy, it has been possible to reduce the length of the waves produced by electric discharges to a few millimetres, so that at the present time the region of infra-red light is separated from that of the waves of electric discharges by a **gap** of only about **three octaves.**

We have seen that the region of electric discharge waves is ranged alongside the region of the infra-red spectrum, with the exception of a comparatively small gap which no doubt will soon be bridged. In recent years still another kind of electromagnetic wave has been recognized as such. The region of these waves lies at the other end of the spectrum, continuous with the ultra-violet. It is occupied by those rays, the discovery of which by **Röntgen** in 1895 aroused world-wide interest.

The question as to the **nature of the Röntgen rays** for a long time remained obscure. Although physicists had grounds for believing that Röntgen rays represent a phenomenon akin to that of light, there was no direct proof of this supposition. Above all, they at first saw no possibility, should their conjecture be correct, of actually determining the wave lengths of the Röntgen rays accurately, since in optics one makes use of a so-called **diffraction grating** for the measurement of wave lengths. A diffraction grating is a metal mirror (preferably concave) on which very fine lines have been scratched by means of a diamond. These lines are exceedingly close together and are equally spaced, the distance between successive lines being of the order of one ten-thousandth of a centimetre. With such diffraction gratings we can measure wave lengths by a very simple method (which we need not describe here), provided they are not very much greater and also not a great deal smaller than the distance between two neighbouring lines on the grating.

Now physicists had reason for supposing that the wave lengths of Röntgen rays, in so far as they resemble light, must be about ten thousand times as small as the wave length of visible violet light. For the determination of the wave lengths of the Röntgen rays one would therefore require diffraction gratings the rulings of which would have to be a thousand times closer than is the case with the finest diffraction gratings that can be produced. One would require gratings the lines of which would be so close together that there would be millions of them in a single millimetre. The technical difficulty of producing such gratings seemed to remove almost every likelihood of physicists ever being able to determine the wave lengths of Röntgen rays. But at this juncture **Laue**, in 1912, conceived the ingenious idea of making use simply of crystals instead of artificial gratings.

Mineralogists regarded the regular arrangement of the molecules and atoms in crystals as the cause of the regular form of such crystals.[18] If this conception is true, then, as Laue recognized, a crystal must give rise to similar effects to a diffraction grating, in consequence of the regular arrangement of its constituent particles. The mode of action would nevertheless be much more complicated, owing to the fact that in a crystal the centres which correspond to the lines are arranged **in space**, and not linearly as in the case of optical gratings. On the other hand, it follows from different physical considerations that the distances between the molecules of a crystal must be of the order of magnitude of less than the millionth part of a millimetre. This would correspond to wave lengths about a thousand times smaller than those of ultra-violet light, and

[18] This theory was due to the mineralogist Bravais, and was developed by him in the middle of the nineteenth century.

would be of the same order as had been suspected in the case of Röntgen rays.

Laue concluded from this that if Röntgen rays are really electromagnetic waves of exceedingly short wave length, they must be diffracted by crystals in quite a similar manner to the diffraction of light by the diffraction gratings used in optics. But there will be differences due to the fact that a crystal is a **spatial** and not a linear grating. If a photographic plate be placed behind a crystal on which Röntgen rays are incident, then by suitable arrangement of apparatus it ought to show on development symmetrically arranged black patches, as foreseen by Laue. Moreover, from the positions of these patches, it should be possible to obtain on the one hand the wave lengths of the Röntgen rays emerging from the crystal, and on the other hand also the structure of the crystal employed. Laue's theoretical considerations were very soon brilliantly confirmed by experiment; the Röntgen rays emerging from the crystal actually did produce on the photographic plate the unique figures he had predicted (Fig. 1).

Laue's discovery of the so-called **interference of Röntgen rays** has become of the greatest importance for the most recent development of natural science. It has revealed to us the **structure of crystals** and thus opened up for crystallography entirely new methods of research. It has led to the discovery of **Röntgen spectra,** which have, within the space of a few years, supplied us with the most valuable information on the internal structure of the atoms. Moreover, Laue's discovery has also brought us the insight into the nature of the Röntgen rays, for which scientists long sought in vain. We shall return to some of these points later.

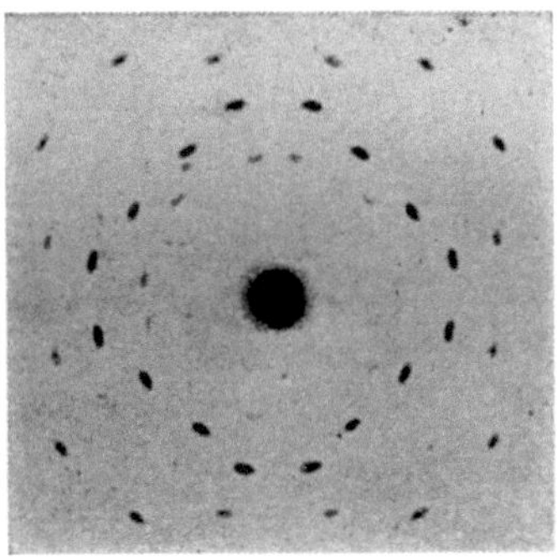

FIG. 1

[To face page 20

The so-called Laue photograms were in themselves a proof of the wave nature of Röntgen rays, and as a result of the measurement of these photograms it was quite clear that Röntgen rays are nothing else than electromagnetic oscillations of particularly short wave length. The rays which are usually grouped together under the name of Röntgen rays comprise a region of some eight octaves,[19] and this begins at approximately seven octaves beyond the visible violet. The gap of about three octaves, which, up to the year 1920, appeared to separate the then known extreme ultra-violet from the Röntgen rays of longest wave length, has now been bridged over by the researches of Millikan already mentioned. In fact, the shortest wave lengths obtained by Millikan and others in the ultra-violet already overlap the longest Röntgen waves. Radioactive substances emit rays of a type similar to Röntgen rays. These are known as **Gamma (γ) rays**, and in the wave scale they are ranged alongside the Röntgen rays of shortest wave length. According to our present-day knowledge they comprise about four octaves so that they extend to about nineteen octaves higher than violet light.

The spectrum visible to the human eye, and embracing the colours of the rainbow from red to violet, thus appears vanishingly small within the complete range of the **spectrum** in its present-day sense. It comprises only a single octave, whereas there is an interval of seven octaves beyond the violet before

[19] In the language of experimental physics the colour of Röntgen rays is termed their "hardness," the basis for this being the penetrating power of the rays. The less the frequency of the rays the softer they are, and conversely.

we reach the region of the Röntgen rays, and one of eleven octaves beyond the red before the Hertzian waves begin. From the point of view of subjective perception the waves of wireless telegraphy, the infra-red heat rays, visible light, the chemically active ultra-violet rays and the Röntgen rays naturally appear as qualitatively different phenomena. From the objective point of view they are one and the same, i.e. electromagnetic waves which are only quantitatively different from each other in their period. In the long run, the possibility and the necessity of these electromagnetic waves are founded, as Maxwell first appreciated, on the two laws that describe the phenomena discovered by Oersted and Faraday, namely, the production of a magnetic field by an electric current, and the production of an induced current by a varying magnetic field.

II. MOLECULAR STATISTICS

THE endeavour to produce a **unified** picture of nature is as old as all theoretical natural science, and even in ancient natural philosophy, this endeavour led to an idea, the extraordinary fruitfulness of which has been clearly evidenced by the newest development of theoretical physics. It was the **atomistic** idea, and originated with the great Grecian philosopher **Democritus.**[1]

The picture of nature received by mankind through his sense organs is not only restricted by the limited power of these organs, but it is also complicated by the **multiplicity** of our senses. Mankind receives sense impressions through the eye, through the ear, and through the temperature sense; moreover, in electricity and magnetism he becomes acquainted with phenomena for which he does not appear to possess any particular sense organ. To this multiplicity of the senses corresponds the traditional division of experimental physics into optics, acoustics, heat, and the science of electricity and magnetism. But it was clear even in ancient natural philosophy that this many-sidedness of the picture of nature can only be of **subjective** origin. For this reason Democritus contrasted this complicated subjective world-picture with a second, **absolute** and **objective** one, in which there was to be only one kind

[1] Democritus was a pupil of Leukippos.

of physical phenomenon, namely, **motion.** But if all physical happening is to be referred to motions, which, owing to the smallness of the moving bodies, will remain hidden from us, then of course **matter** itself would have to be regarded as being entirely **resolvable** into invisible small particles.

The further development of physics has nevertheless shown that the complete reduction of the whole of physics to mechanics, for which Democritus hoped, cannot be effected. None the less, what was to Democritus only a bold speculative hypothesis has become a certainty as a result of modern physics, viz. that matter is made up of very small and very quickly moving particles. Moreover, during the last half century it has become generally recognized in physics that just these invisible motions of the material particles are responsible for the phenomena of **heat.**

Although the first beginnings of the **kinetic theory of heat** date back to the seventeenth century,[2] it was not until the middle of the nineteenth century that its exact development began, and this in intimate connection with the discovery of the so-called **first law of thermodynamics.** In the year 1842 Robert **Mayer** and **Joule** made the very important discovery that in the production of heat by mechanical work and also in the converse process, there is a **constant transformation ratio** between the quantities of **heat** and of **mechanical energy** mutually converted.[3] This

[2] A mechanical interpretation of heat is to be found in the work of Roger Bacon; at the beginning of the nineteenth century the mechanical theory of heat was re-founded by Rumford, who supported it by important experiments.

[3] This transformation ratio is called the mechanical equivalent of heat, or the calorific equivalent of work. A quantity of heat of

remarkable fact was capable of an extremely simple explanation when one considered heat to be nothing else than motion, for the law of the **conservation of mechanical energy** in true processes of motion had long been recognized as a necessary mathematical consequence of Newton's laws of motion.[4] The total mechanical energy involved in the production of heat by work would, according to the kinetic theory of heat and in the sense of mechanics, thus remain as such unaltered. Solely the distribution of the total amount of mechanical energy between two part-amounts would vary; the first of these being the part that appears to us to be necessitated by the motion of the visible bodies as a whole, and the second being one which we ascribe to the invisible motion of the smallest particles of which the bodies are composed. Since the Greek for large is *makros*, and for small *mikros*, we can distinguish between the two types of motion and thus between the two part-amounts of energy by using the terms macro-mechanical and micro-mechanical.[5]

one calorie, which would heat 1 gram of water from 15° to 16° C., is equivalent to the work done when a mass of 427 grams weight is raised through 1 metre.

[4] The law of the conservation of mechanical energy presumes, however, that the effective forces have particular properties. It is fulfilled only when such forces are effective between the bodies of the system as act along the direction of the connecting lines and which only depend on the distance. These conditions are fulfilled both for gravitational forces and also for electrical and magnetic forces. The law of the conservation of mechanical energy is to be found in precise form in the works of Lagrange even in the eighteenth century. Helmholtz was the first to show in an exact manner (in 1847) how the law can be applied also outside the realm of mechanics (in the narrower sense of the word).

[5] The beginnings of the kinetic theory of gases go back as far as the year 1738, when Daniel Bernoulli published the fundamental notions of this theory in his "Hydrodynamics."

The atomistic principle first showed its extraordinary fruitfulness in the field of the **kinetic theory of gases,** which was founded in an exact form in the sixth decade of the nineteenth century by **Krönig** and by **Clausius.**[5] The basal idea of the kinetic theory of gases is that for gases the motion of the smallest particles, the so-called **gas molecules,** is simply one of **translation,** and the molecules are not bound to particular positions of equilibrium, as we assume to be the case for the solid and liquid states.[6] From the atomistic standpoint we must thus regard a gas as being a swarm of small particles shooting about in all directions, their motion being influenced only by the presence of the other molecules, near to some of which they approach in their flight.[7]

The essential properties of gases can be deduced immediately and in the simplest manner from this fundamental hypothesis. The **pressure** exerted by a gas on the sides of the vessel in which it is enclosed

[6] This conception is due to Joule (1851).

[7] The classical theory of gases compares gas molecules with elastic spheres which are constantly undergoing collision in consequence of their finite magnitude. The number of collisions to which a molecule would be subject in unit time on this assumption can be calculated from the internal friction (viscosity). At normal temperature and pressure it would amount to several milliards in a second. According to the modern conception, which, however, has not yet been developed in an exact manner in its application to the theory of gases, there would no longer be any real point in speaking of collisions. The molecules would have to be compared with tiny planetary systems, from the constituents of which electrical forces emanate (see the fourth lecture). Electrical forces would thus act between the molecules and constantly deflect them from their rectilinear paths, the more so the nearer two molecules approached each other. The motion of the molecules would thus take place (without sharp corners) in a zig-zag fashion.

can be very simply explained by the unceasing **impacts**, the constant collisions of the gas molecules with the sides of the vessel. If the density of a gas be doubled, there will then be twice as many impacts per second on a given area of the vessel walls, and thus, under otherwise equal conditions, the pressure must be proportional to the density of the gas, or inversely proportional to the volume, which amounts to the same thing. In point of fact, this is the fundamental relation in the study of gases, and was recognized as such by physicists in the middle of the seventeenth century.[8]

The **temperature** of a gas is determined by the mean value of the energy of motion falling to the lot of a single molecule. On the other hand, the value of the energy of motion itself depends on the molecular velocity, and is proportional to the square of this quantity. The temperature at which the molecules of such a gas would have absolutely no energy of motion, i.e. would be in a **state of complete rest**, is called the **absolute zero of temperature.** This idea had of course been familar to heat specialists previously, and independently of the kinetic hypothesis; the absolute zero was understood to be the temperature at which the pressure of all gases disappears. From the observed dependence of pressure on temperature it follows that the absolute zero must be situated 273·1° below the melting-point of ice on the Centigrade scale. In passing, it may be mentioned in this connection that the lowest temperature hitherto produced experimentally in

[8] This regularity was recognized by Boyle in the year 1660 and is hence called Boyle's law (it is also falsely known as Mariotte's law).

the laboratory is $-272{\cdot}2°$ C., which lies only $0{\cdot}9°$ above the absolute zero.[9]

More detailed purely deductive investigation has shown that, for a definite temperature, the **molecular velocity** can be stated without further trouble, for we know from empirical measurements how the pressure and the volume of a gas vary with the temperature. Clausius showed in this way that molecular velocities are of the same order of magnitude as the velocities with which projectiles are fired from guns. For hydrogen at a temperature of 0° C., for instance, the average velocity of the molecules amounts to 1800 metres per second.[10] In the case of other gases it is not so large, being smaller the greater the molecular weight of the gas. For the two main constituents of the air, oxygen and nitrogen, for instance, the velocity is only about one quarter of that for hydrogen.

Apart from Clausius, the kinetic theory of gases was further developed first by Maxwell, the creator of the electromagnetic theory of light.[11] But probably the most interesting result to which the further investigations on the kinetic theory of gases led was due to the Viennese physicist **Loschmidt**, since he was the first to determine the **absolute**

[9] This temperature was produced artificially by Kamerlingh Onnes by the use of Linde's regenerative circulation apparatus and by means of liquid helium. At this temperature (0·9° A.), helium is still in the liquid state.

[10] Even before Clausius, Joule had calculated molecular velocities, although his values were inexact and also false.

[11] Maxwell's researches had to do especially with the viscosity of gases, and with the law according to which different values of the velocity are distributed amongst the molecules of a gas. The more the values of the velocity deviate from the average value, the less frequently they occur.

magnitude of the molecules.[12] He succeeded in this in the year 1865 as a result of considerations which were concerned on the one hand with the viscosity of gases, and on the other, with the liquifaction of gases. As we shall see in greater detail in a subsequent lecture, theoretical physics has found since then various much more exact methods whereby to determine the true magnitude of the molecules. We can even say that to-day we know the value for the mass of the molecule of hydrogen with comparatively greater accuracy than, say, the value for the mass of the earth.

The exact value which we are now able to give for the mass of a molecule of hydrogen indicates that a **quadrillion (i.e. a billion billions) molecules of hydrogen** only possesses a **mass of 3 grams.**[13] Since the mass of the earth is estimated at about 6000 quadrillions of grams, it follows that the mass of a molecule of hydrogen would be related to that of a stone of mass about 130 grams in approximately the same way as is the mass of the stone to that of the whole earth.

We thus see how exceedingly small are the carriers of the invisible motions that appear in experimental physics as heat. The processes, the conditions, which are single processes and conditions to the experimental physicist, thus appear to the atomistic physicist as a **totality** of a vast number of **individual** processes and of individual conditions; moreover, their carriers are those material

[12] The relative magnitudes of the molecules, i.e. the ratio of the masses of different molecules to the mass of an atom of hydrogen, had, of course, been long known in chemistry.

[13] The mass of the molecule of hydrogen amounts to $3{\cdot}2 \cdot 10^{-24}$ grams.

individuals, the weight of which does not in general amount to even the trillionth part of a milligram.

One and the same physical condition can thus be regarded from two quite different points of view. If, for instance, we have a gas in a large vessel fitted with a movable piston, the experimental physicist knows all that he desires to know about the condition of the gas, when he knows the pressure exerted on the sides of the vessel by the gas, the temperature of the gas, and the volume occupied by its mass.[14] To him, these data complete everything that appears to be worth knowing about the condition of the gas. To the atomistic physicist, on the other hand, the gas is made up of quadrillions of molecules, and hence he regards the condition of the gas as being resolved into quadrillions of individual, albeit purely mechanical conditions. His knowledge of them at a particular moment would not be complete until be knew for every single molecule what position it occupies, and with what velocity and in what direction it is moving at the instant under consideration. We are led to ask the question as to whether atomistic physics, instead of fulfilling its purpose of simplifying our physical conception of nature, does not in reality bring about an immense complication of that conception. We shall find the answer to this question in a simple consideration made in a domain which is quite apart from physics.

[14] Between the three so-called magnitudes of condition, pressure, volume, and temperature, however, there is an equation, the so-called equation of state, which connects these three magnitudes together. It expresses Boyle's law, or, more generally, Van der Waals' law. If we know two of the magnitudes of condition, we can always find the third.

In social economics we speak of age distribution of the population, of birth, marriage, and mortality rates, of the percentage of suicides and of similar conceptions, and we speak of laws demonstrated by these magnitudes; and yet the population, which as a whole can be the subject of such socio-economic research, is made up of a vast multitude of individuals, the life of each one of which is undoubtedly characterized by chance and irregularity, particularly in regard to such voluntary actions as marriage or suicide.

It is found that, in so far as we take account of a sufficiently large number of individual cases, the individual factor falls quite into the background. The part played by individual fortuities is smaller the greater the number of the individual cases which form the subject of our consideration. This remarkable fact, which is known as the **law of large numbers**, constitutes the basis of all **statistical** investigations.

An illustration may serve to show this more clearly. In a city with a population to be counted in millions, the number of deaths per week is represented by a relatively large number (about 500 for every million). Now we find that in such a city the weekly mortality only varies slightly, relatively speaking, in the course of a month. In fact, individual fortuitousness becomes so imperceptible that, were we to find appreciably different numbers for the weekly mortality for two weeks separated by a somewhat long interval of time, we should seek for some outside cause of this. Perhaps more people have died during a week in winter than during one in summer, in consequence of the weather, or the number of cases of death may have been increased by an epidemic or famine.

We can only regard outside causes as being responsible for an alteration of such a number as the weekly death-rate when this number is sufficiently large. Instead of a large city, suppose we consider a small town of about 6000 inhabitants, in which on the average about three persons die weekly. If the number of deaths in the first week of March in one year were four, and two in the corresponding week of the following year, we should be quite wrong were we to infer from this chance occurrence that the conditions of health had improved in that town within the course of the year. For such **small numbers** as four or two the **individual factor** plays an important part; for such small numbers we may have relatively important alterations in their values, and considerable **fluctuations** may occur **without external cause,** solely in consequence of individual hazard.

In consequence of their general validity, these simple considerations on population statistics can be applied straightway to the study of molecular phenomena. Just as the magnitudes that play a part in population statistics depend on individual magnitudes which characterize individual conditions or processes, so this must also hold for the magnitudes which, conformably with the kinetic theory, occur in the study of heat. If the physical process under consideration is of such a kind that the number of the individual processes into which it appears to be resolved from the atomistic point of view is **large,** then the comparison with the city of millions is applicable. By **addition** or by **averaging** the individual values we can then derive **statistical magnitudes** which have reference to the **whole as such,** in just the same way as in population statistics we count up the individual cases of death, or calculate an average duration of life from the ages attained by the deceased individuals. Moreover, if the comparison with the city of millions is permissible, then it will also be true for physical statistical magnitudes that they do not vary appreciably in the absence of an outside cause, in so far as they depend on **a large number of individual magnitudes.**

An illustration will make this clearer. We imagine a vessel filled with a gas, and within it a very small cube with edges only a ten-thousandth part of a millimetre in length. In consequence of the internal state of motion of the gas, molecules will be continuously darting, with tremendous speed, through this cube. We shall suppose a fictitious observer, with the ability to recognize single molecules as such, to undertake at arbitrary instants an instantaneous count of the molecules momentarily situated within this imaginary cube. If this fictitious atomist were to multiply the number thus obtained by the mass of a molecule, all of which are of the same size, and then divide this quantity by the volume of the cube, he would obtain a magnitude which would be identical with that designated in descriptive physics as the **density** of the gas at the place in point.

The fictitious observer might also measure with lightning rapidity the velocities of the single molecules at a particular moment. Were he to multiply the mass of a molecule by half the square of the molecule's velocity, he would obtain the energy of motion of that particular molecule. From the values so obtained for the energies of motion of all the molecules momentarily situated in the small cube, he could then find the average energy of a molecule. Moreover, in consequence of a simple relation, this average energy would give a determination of the magnitude known to the descriptive physicist as the **temperature**, at the place where the imaginary small cube was supposed to have been constructed.

Although this cube was supposed to be very small, with an edge one ten-thousandth of a millimetre in length, it follows from what was previously stated about the size of molecules, that even then it still contains in general thousands of molecules.[15] Thus we may still apply the comparison with the city of millions, and this in spite of the exceeding smallness of the cube the volume of which was involved in the statistical considerations. The individual molecular magnitudes in themselves are thus usually quite of secondary importance to the conditions and processes described in experimental physics. As far as the descriptive physicist is concerned, only such values are important as are obtained by statistical methods from a large number of single values which of themselves are of little value. But in consequence of the large number of individual processes, the spontaneous variations of these statistical magnitudes will be relatively very small; as a result of this, no fluctuations of appreciable magnitude will occur in general in the density or temperature of a gas, without outside causes. Hence, a thermometer suspended in a room will not show a sudden rise or fall in temperature without the intervention of some external cause. Conversely, moreover, if such a variation of the reading of a thermometer does

[15] According to Avogadro's law (1811) all gases at the same pressure and temperature contain the same number of molecules in a given volume. Referred to atmospheric pressure and to a temperature of 0° C., the number of molecules in 1 c.c. of a gas is about 28 trillions.

occur without our knowing the cause, we should necessarily ascribe this alteration to an outside cause unknown to us.

Let us return to our atomist who is able to recognize the individual molecules, and suppose that he now restricts his statistical investigations to a still smaller cube. He isolates an imaginary cube the edges of which are of length only one two-hundredthousandth of a millimetre, i.e. twenty times smaller than before. At ordinary pressure and temperature, there will be on the average only three or four molecules present at any instant in this very small volume. In this case the comparison with the small town will be applicable. If the atomist determines the density at the place under consideration from such a small number, the density he obtains will reveal relatively large fluctuations solely in consequence of the individual irregularities, and without external cause.

As to whether or not a physicist detects **fluctuations in density** will thus depend solely on the smallness of the volume chosen for his molecular statistical considerations. From an objective point of view, molecular fluctuations of density will be unimportant for all processes in which distances of about a hundredthousandth of a millimetre are too small to be of moment for the process. On the other hand, for those processes in which such small (or not appreciably greater) distances have to be taken into account, the effect of the molecular fluctuations must make themselves manifest. The passage of a ray of light through a gas belongs to this class of phenomena. The value of the quantity that characterizes the optical behaviour of a gas, i.e. the so-called refractive index, depends on the density of the gas. Fluctuations in the density must therefore result in **fluctuations in the value of the refractive index.**

The great progress made in theoretical atomistics since the beginning of the twentieth century is due largely to **Smoluchowski.** He recognized that such fluctuations of the refractive index in the atmosphere are responsible for the **blue colour of the sky.** The correctness of this interpretation has been established without doubt by different interesting experiments.[16] In fact, it was possible on the basis of Smoluchowski's theory to calculate the size of the molecules from optical measurements on the colour of the sky, and the results are in good agreement with those obtained by quite different methods. The phenomenon of the blue of the sky has been known since the beginnings of mankind, and yet we are indebted to modern physics for recognizing in it a brilliant empirical confirmation of the atomistic hypothesis, devised long ago on purely philosophical grounds by the genius of a Grecian thinker.

A still clearer proof of the correctness of the kinetic molecular theory is to be found in a phenomenon first observed about a hundred years ago by the botanist **Brown,** and generally called after him **Brownian motion.** An object suspended in the air will, according to the atomistic conception, be subject to collisions on all sides due to the molecules of the air striking against it. But if the object is so large as to be visible to the naked eye, it will experience many millions of collisions even in

[16] These experiments have reference to the so-called phenomenon of opalescence.

vanishingly small fractions of a second, in consequence of the immense number of the molecules. Since these collisions take place from all possible directions, they will almost completely counterbalance each other in their effects, owing to the regularity brought about by the large number. The result of this will be that, although the suspended body will assume a sort of quivering motion from the impacts of the molecules, the magnitude of this will nevertheless be much too insignificant for it to manifest itself in any way in the case of objects visible to the naked eye. On the other hand, suppose we consider a much smaller object, say, a suspended particle of matter so small that it is not visible in a microscope, and can only be recognized in an ultra-microscope.[17] In this case the individual irregularities must reveal themselves in a noticeable manner, and the particle must perform a vigorous zig-zag motion as a result of the irregular collisions on all sides.

As a matter of fact, in the year 1827, Brown discovered remarkable irregular motions under the microscope, which occurred in liquids containing plant pollen, when the linear dimensions of the pollen were of the order of a two-hundredth part of a millimetre. Furthermore, Brown also showed

[17] The ultra-microscope invented in 1903, by Siedentopf and Zsigmondy, rendered possible the detection of particles that cannot be seen in an ordinary microscope because they are smaller than the wave lengths of visible light. Ultra-microscopic particles, which are thus still smaller than about a thousandth of a millimetre, can be recognized by means of an ultra-microscope in virtue of the fact that they diffract the light that passes them ; but they are not themselves visible as regards their shape. In consequence of this the particles appear like small luminous stars, but their form remains unrecognizable.

that the smaller the pollen the more lively are these motions. Half a century later, Christian **Wiener** first recognized that the true cause of this phenomenon lay in the internal state of motion which must be ascribed to the liquid, according to the kinetic theory of matter. The existence of the Brownian movement in the case of **gases** was not proved until much later, by **Ehrenhaft**. The exact development of the **theory** of the Brownian movement is primarily the work of **Einstein** and of **Smoluchowski.**

Our previous discussion shows clearly that in the long run all molecular statistical investigations are founded on **probability considerations**. We have seen that relative fluctuations are greater the smaller the number of the single values from which the fluctuating value has been formed, and that the fluctuations become unnoticeable when the number of these single values becomes large. This is explained by the fact that the probability of a variation of definite relative amount (say of a fortuitous doubling) decreases rapidly, even when the number of the single values is only increased from three to twenty. In a small town in which on the average about three persons die weekly, it may be that on one occasion perchance nobody died in one week. This need by no means be a rare occurrence. For a city of millions, on the other hand, such an event is so vastly improbable and hence so exceedingly rare, that we can confidently assert that since first such cities existed and no matter how long they may exist, such a case has never yet arisen, and, moreover, never will arise. We can make this confident assertion because the probability that it is erroneous is inconceivably small. An event the probability of which is exceedingly small we designate as being **impossible.** The difference between the two conceptions is so slight that in general differentiation appears to be superfluous. We shall make this clearer by means of illustrations from quite another domain than that of actual physics.

In a certain town there is a hall in which daily lectures are held. For the purpose of simplicity we shall assume that the number of people in the audience is the same for each lecture. We shall ask ourselves the question as to whether it could happen that, **purely by chance**, only such persons are present on a particular day whose names begin with the initial letter M. In what follows we shall call this event briefly an M-event. Anyone who does not ponder long over the matter would certainly straightway deny the purely fortuitous possibility of such an M-event, but the complete answer is not given without the following simple consideration. We have as yet supposed the size of the audience which is the same day by day to be quite arbitrary. Let us assume that the constant number of listeners is only two, and further, that about one in twenty of the inhabitants of the town have surnames beginning with M. Then the probability of the M-event for an audience of two is one in 400, or in other words, from amongst every 400 lectures there will be an average of one in which the M-event occurs. Thus if the lectures take place daily, the M-event will recur on the average every 400 days, and for this reason the period of 400 days is called the average **recurrence interval** (period) of the M-event for two listeners.

If the number of people in the audience is three instead of two, the recurrence interval will be twenty times as large; for each increase of one in the number of people, the period always becomes twenty times greater in consequence. For five people it would already amount to 8000 years, for seven people about 3 million years, for ten people about 30 milliards of years, and for twenty people about a third of a quadrillion years. For 100 people the recurrence interval of the M-event would be given by a number of years represented by more than one hundred figures. Even if the town, the inhabitants of which have names, should be as old as the earth or the solar system, and even if lectures had been delivered daily to 100 people throughout the whole of this time, even then the probability would be exceedingly small that in this inconceivably long period an occurrence would have arisen even once, in which by pure chance only people with the initial letter M had been present at a lecture.[18]

From the ordinary human standpoint it would be meaningless to reckon with still greater periods of time, and we shall therefore describe the purely fortuitous occurrence of the M-event as being impossible. From the anthropomorphic (human) point of view, which of course is not an absolute one, we regard quite generally those events as **impossible by chance**, the **recurrence interval** of which is **large beyond the limits of human conception.** For the same reason we shall also regard it as impossible that purely by chance the people in a lecture-hall should sit so that

[18] Of course the statistical probability of the M-event could also be increased by outside causes; there might be, for instance, a philologico-historical lecture on the origin of the surname Miller, and, as we can readily understand, this lecture would be of particular interest to those persons bearing this name with the initial letter M.

on the right-hand side of the hall there are only persons the initial letters of whose names range from A to K, and on the left-hand side only those with initial letters L to Z.

Owing to their general nature, the foregoing considerations can be applied straightway to molecules as well as to people. Just as people's names have different initial letters, so the molecules have different directions of motion and different velocities.[19] From the standpoint of molecular statistics, a process known to experimental physics as a process of motion (in the narrower sense of the word) would be understood as one in which a tremendously large number of molecules is moving in the same direction and with the same velocity. If this be the case, then in the sense of experimental physics we speak of the motions of a whole body, or of currents or vortex motions which take place in the interior of liquids or of gases, and we are then dealing with processes that possess macro-mechanical energy as contrasted with micro-mechanical heat energy. The event that at a particular moment 100 neighbouring molecules should chance to have the same motion would be comparable with the M-event previously discussed. The former event, like the latter, we should have to designate as impossible by chance, from the anthropomorphic point of view. In fact, in a gas, which appears at rest to the experimental physicist, the probability that **a "tangible" mechanical process should arise by chance from the thermal motion** is still inconceivably smaller than the probability of the M-event with 100 individuals, because for the production of a macro-mechanical

[19] Compare note 11.

process not 100 but many many millions of molecules must be involved.

In their irregular motion which we call heat, the molecules proceed not only in different directions, but in addition, as already mentioned, they have velocities of different magnitudes. Now imagine a vessel in which a gas is contained. If it could happen by chance that the right-hand side of the vessel were to be occupied mainly by those molecules with velocities greater than the average, and the left-hand side by those with velocities smaller than the average, this would be equivalent to the chance occurrence of a difference of temperature, inasmuch as the right-hand side of the vessel would be warmer and the left-hand side cooler. Such an event appears to be just as impossible in practice as that previously given, in which an arranged distribution of the audience in the two halves of the lecture hall was supposed to arise by chance. Hence from the point of view of mankind, but only from this point of view, we shall have to regard it as impossible that heat energy is transformed of itself into concrete mechanical energy by chance, or that temperature differences can arise of themselves within a uniformly warm body.

Quite generally speaking, however, if macroscopic or gross conditions of greater probability are more frequent than those of lesser probability, then, of course, every gross condition of small probability must show an **apparent tendency to pass over into a gross condition of greater probability.** From his viewpoint, the experimental physicist must therefore recognize that **macro-mechanical energy** shows a **tendency to transformation into thermal energy,** that every macro-mechanical process in consequence of friction must always be connected with a **development of heat**, and that on the other hand, existing **differences of temperature** must have the tendency to **equalize** each other. Thus, since the chance transformation of macro-mechanical energy into thermal energy has a probability amounting to complete regularity, whilst the converse process has a probability small beyond the possibility of conception, and since the same thing also holds for the equalization of temperatures and the converse process, all natural occurrences must therefore appear to be **irreversible** to anthropomorphic physics.

This fact forms the content of a law which was introduced into the study of heat in the middle of the nineteenth century, and called the **second law of thermodynamics.** Owing to the philosophical outlook that it appeared to open up, this law always had a certain mystical character. From the point of view of modern molecular statistics the second law is not so much a law as a rule, for the fulfilment of which there is nevertheless always a probability practically equivalent to certainty. Thus the second law can only be regarded as true by a physics which, in the sense of the words of a Grecian thinker,[20] "applies to all things the scale of mankind." This law can only be designated true by a physics that calls every assertion true, for which the probability of disproof by experience is smaller than can be conceived by the human mind. But the second law must no longer claim complete validity in regard to such applications as go beyond human comprehension.

This new interpretation of the second law is due

[20] It was Protagoras.

to **Boltzmann**, who first found an atomistic meaning for the law in the year 1866, when he reduced it to probability considerations.[21] The clear investigation and completion of Boltzmann's ingenious idea again falls to the credit of **Smoluchowski**, however, whose researches took place in the first two decades of the twentieth century.

As we have seen, the human mind is only able to comprehend a **regulating effect of chance** when the number of the individual events is small. The illustration of our so-called M-event showed us that even with a number of only 100 individuals (who, so to speak, belonged to twenty species), the human mind is not in the slightest degree capable of comprehending any actual order as being the work of chance. Many quadrillions of individual processes take place in the air of a room owing to the extraordinary smallness of the molecules. In the case of systems with such a tremendous number of individuals, if we wished to state the recurrence interval of a chance arrangement sufficiently large to be detectable by the senses, the number of years which determines the period would be inconceivably large. It would be so vast that if we wished to write it down on a strip of paper stretched between the earth and the moon, we should be a long way from having completed our task even when we had filled the strip.

After what has just been said, how could the human mind comprehend the possibility that the

[21] Entropy plays an important part in thermodynamics. In the case of reversible processes it remains unaltered, but apart from these it always increases. Boltzmann recognized that the entropy is identical with the logarithm of the statistical probability, except for a constant proportionality factor.

order of the universe could have arisen by chance from the chaos of the irregular molecular thermal motion of our universe? But the incapacity of our mind to grasp it cannot constitute a proof against such a possibility. Not physics, but only philosophy can judge as to the justification of such an assumption. Modern molecular statistics has demonstrated the theoretical possibility of such an assumption, albeit only in connection with its **incomprehensibility**.

III. THE ELECTRON THEORY

THE foundation of the electromagnetic theory of light and the creation of the mechanical theory of heat represent the two great accomplishments of theoretical physics in the second half of the nineteenth century, and both were the product of an endeavour to unify our picture of nature. In consequence of Maxwell's theory, **electricity** assumed the leading rôle amongst the forces of nature. On the other hand, the mechanical theory of heat owed its great successes to the principle of **individualization** in its application to matter. Hence the further development of physics rendered it imperative that the principle of individualization, which had already been recognized as so fruitful, should also find an application in the science of electricity. This was obviously only possible by the assumption of **small individual electrical charges,** which had to be brought into some connection with the individuals known as **molecules,** into which the mechanical theory of heat **resolved matter.**

On the other hand, the fundamental facts of **chemistry** had for a long time forced us to the conclusion that molecules are built up from a number (in general not large) of so-called **atoms,** of which there must be just as many kinds as there are **elements** in chemistry. This hypothesis was formulated by **Dalton** in the year 1805. It is based on an empirical law discovered by Dalton, and known as the **law of multiple proportions**. This law can be expressed in the following form: A particular number, characteristic of the element, can be allotted to every chemical element, in such a way that the quantities of the elements contained in a chemical compound are related to each other as whole number multiples of the numbers characteristic of the elements concerned.

The law of multiple proportions receives a very simple interpretation by the assumption that the numbers characteristic of the elements represent nothing else than the **relative weights of their atoms**, and that the molecules themselves are built up from the atoms of the elements in chemical combination. For instance, we regard a molecule of water as being composed of two atoms of hydrogen and one atom of oxygen, the atomic weight of the latter being sixteen with respect to hydrogen;[1] i.e. an atom of oxygen is about sixteen times as heavy as an atom of hydrogen. A molecule of sulphuric acid is regarded as made up of two atoms of hydrogen, one atom of sulphur, and four atoms of oxygen, and similarly for other chemical compounds.

Whilst the Dalton theory considered the atoms of the chemical elements to be the bricks of matter, the fundamental facts of **electrochemistry**, on the other hand, gradually led to the conclusion that **definite electrical charges** must be **combined with the**

[1] Relatively to hydrogen, the exact atomic weight of oxygen is 15·88. Usually, however, the unit of relative atomic weight chosen is not hydrogen, but the sixteenth part of the atomic weight of oxygen, which is only slightly different from the other unit.

atoms. In the year 1833 **Faraday** had discovered exceedingly simple laws for the process of **electrolysis**, i.e. for the chemical decomposition of liquid conductors of electricity by means of an electric current. The best known example of this is the decomposition of water into its components hydrogen and oxygen, the former of which is evolved where the current leaves the liquid, and the latter where it enters.[2]

The phenomenon of electrolysis can be very simply explained if we assume that in liquid conductors of electricity the molecules are entirely or in part split up into two oppositely charged components, the so-called **ions**.[3] For instance, we might assume that a molecule of water is broken up into a positive electrical hydrogen ion consisting of two hydrogen atoms, and into a negative electrical oxygen ion consisting of one oxygen atom. The conduction of electricity in liquids must then be regarded simply as a **transport** of such ions, which wander in opposite directions in consequence of their opposite electrical charges, and in such a way that the positive ions are liberated at the place

[2] The direction of the current is designated as that in which the positive electricity flows, or, what is the same thing, the direction opposite to that in which the negative electricity flows. Since it is quite arbitrary and purely conventional that we call the so-called glass-electricity positive and the so-called resin-electricity negative, rather than conversely, it is of course also purely convention as to which of the two " electrodes " we call the place of entrance (anode) of the current and which we call the place of exit or cathode. Moreover, when water is made slightly acid to increase its otherwise small conductivity, the electrolytic decomposition of water constitutes only a secondary effect; the primary effect consists in the decomposition of the acid.

[3] In Greek the word *ion* signifies "the traveller."

where the current leaves the liquid and the negative ones at the place where it enters.[4]

The laws of electrolysis[5] discovered by Faraday can be very simply interpreted by the assumption that each atom of hydrogen in its electrically charged or so-called **ionized** state is always associated with exactly the same quantity of electricity. This quantity is now sometimes called the **elementary quantum of electricity**, or the **fundamental charge**. We must further assume that the charge of any ion whatsoever is equal to the fundamental charge, is twice as large, or three times as large, and so on, according to the so-called chemical valency of the ion under consideration.[6]

The magnitude of the elementary quantum is now known very accurately by numerous methods of which we shall speak in the fourth lecture. The so-called absolute electrostatic unit of quantity in electricity is made up of about two milliards of elementary electrical quanta. We can get some idea of the magnitude of the absolute electrostatic unit when we consider that the usual charges of Leyden jars amount to only a few electrostatic units.[7]

[4] This conception originated with Grotthus (1805); its real development was the work of Hittorf and Kohlrausch in the fifties of the nineteenth century.

[5] According to the electrochemical laws of Faraday, the amount of a substance electrolytically decomposed in a given time is solely determined by the current, to which it is proportional. On the other hand, the amounts of the constituents deposited from different electrolytes by a current of a particular strength are chemically equivalent, i.e. they are related to each other as the quotients of the atomic weight and the chemical valency.

[6] Compare the fifth lecture.

[7] The absolute electrostatic unit of electricity is defined as that quantity of electricity which exerts a force of 1 dyne on an exactly

The charge of an ionized atom of hydrogen is thus not large absolutely, although it is relatively very considerable owing to the exceeding smallness of molecules and atoms. In the case of two ionized hydrogen atoms, two different forces act between them. In the first place, there is the force of gravitation, since the two atoms gravitate towards each other owing to their masses; secondly, there is an electrical force, because the two electrical charges attached to the ionized atoms exert on each other a force of repulsion. Both of these forces are inversely proportional to the square of the distance between the atoms. By means of a simple calculation we can easily show that the electrical force between the ionized atoms is about a sextillion times as large as the gravitational force between them, i.e. about a trillion trillion times as intense.[8]

The conclusion that the atoms must be combined with electrical charges gained quite a definite significance as a result of two important new conceptions which were introduced into the theory of electricity about the year 1880. These are the conceptions of the convection current and of electromagnetic mass. The idea of a **convection current** re-

equal quantity placed at a distance of 1 cm. A dyne is the force that imparts to a mass of 1 grm. an increase in velocity of 1 cm./sec. in each second of time. A dyne is slightly larger than the weight of 1 mg.

[8] The mass of an atom of hydrogen is equal to $1 \cdot 6 . 10^{-24}$ grm. (cf. note 13 in the second lecture). The force exerted on each other by the atoms at a distance of 1 cm. is obtained in dynes by raising this number to the second power, and multiplying this by the universal constant of gravitation, the value of which is $6 \cdot 68 . 10^{-8}$ absolute units. The force exerted between the charges is obtained simply by squaring the number for the elementary electrical quantum (fundamental charge) in absolute units, viz. $4 \cdot 77 . 10^{-10}$ e.s.u.

sulted from the investigation of the effects produced by a **moving electrical charge.** It was suggested that the motion of a charged body represents an electric current, and that such a convection current has the **same properties** as had been determined empirically in the case of electric **currents in wires.** An experiment of the American physicist **Rowland** in the year 1876 confirmed that this supposition is actually fulfilled. Rowland set a plate covered with electrically charged tinfoil sectors in rapid rotation, and was thus able to deflect a neighbouring magnetic needle in exactly the same way as would have been brought about by an electric current.

From the fact that a convection current possesses just the same properties as an electric current, Joseph John **Thomson** in the year 1881 drew a surprising conclusion. According to Maxwell's theory of electricity and magnetism, a certain amount of **energy** must be associated with every electric field or magnetic field as such. On the other hand, a moving electric charge in virtue of its motion produces a **magnetic field,** since it represents a convection current. This magnetic field and hence also its energy are only present when the charge is in motion. In accordance with the law of the conservation of energy, however, it is impossible that this energy of a magnetic field should be produced from nothing. An **expenditure of energy** is thus unquestionably necessary in order that an electric charge may be set into motion. Of itself, then, an electric charge possesses the property known as inertia or **inertial mass.** For example, we know that a stone possesses an inertial mass, because this is expressed in the fact that an expenditure of energy is necessary in order to set it in motion, or to give to it a definite velocity. Thus every electric charge behaves as if it had a definite mass, and this is called its **electromagnetic mass.** Closer investigation shows that the electromagnetic mass must be proportional to the square of the charge, and if the charge is in the form of a sphere, the electromagnetic mass must also be inversely proportional to the radius of the sphere.[9] Hence the smaller the radius, the larger will be the mass.

We now see that the idea of mass is absolutely inseparable from the idea of electrical charge. From this it follows that **every convection current** must **simultaneously represent a mechanical process.** Every convection current can thus be considered from two quite different points of view; both the laws of electricity and those of mechanics can be applied to it. It is just this **simultaneous applicability of both kinds of laws** to the same process, or, in other words, this electro-mechanical parallelism which gives rise to a series of new relations, the great fruitfulness of which became obvious when, in the year 1895, the Dutch physicist **Lorentz** produced the **electron theory.**[10]

The theory of Lorentz is based on the idea of the convection current. It assumes that small electrical charges capable of motion are contained in the molecules, and that the alterations in the positions of these represent convection currents. On the basis of the electrochemical laws it further assumes that these charges, which Lorentz calls **electrons,**

[9] When the charge resides wholly on the surface, we must multiply the quotient of the square of the charge and the radius by 2/3, and divide this by the square of the velocity of light, in order to obtain the actual electromagnetic mass.

[10] The beginnings of Lorentz' theory go back to the year 1880.

are each of magnitude one elementary electrical quantum.

The theory of Lorentz represents an **extension of the theory of Maxwell.** The essential and well-established foundations of Maxwell's theory are retained; but by the introduction of the electron hypothesis, the electromagnetic theory was now also able to explain several phenomena where the original Maxwell theory had completely failed. The theory of Maxwell had not only disregarded the electrochemical laws discovered by Faraday, but also the phenomena of the discharge of electricity in rarified gases. Above all, there was a fact that had long been known in optics, which undoubtedly demanded the further development of Maxwell's theory. This was the phenomenon of the **dispersion of light,** i.e. the separation of light into its constituent colours.

Newton was the first to go into this question at all thoroughly. He had already made the important discovery that light of different periodicity is refracted by different amounts. However, the explanation of this fact offered the greatest difficulties both to the elasticity theory of light and to the Maxwell theory in its original form. In contrast to this, a simple explanation was forthcoming when it was assumed that electrically charged particles possessing inertial mass are contained in the interior of the molecules, and that these particles carry out mechanical processes of a periodic nature. If such an electrically charged particle (an electron in the sense of Lorentz' theory) be struck by a light wave, it will be acted upon by a force, of which the direction or magnitude or both vary periodically, since a wave of light is nothing else than an electric field varying periodically in space and time. A periodically varying force is thus acting on the electron struck by a wave of light, whereas the electron itself is carrying out a periodic motion. This superposition of two periodic processes must therefore lead to phenomena such as those called **resonance** in acoustics, or more generally termed **forced vibrations**.[11] A theory of dispersion was founded on the basis of these ideas, but we cannot deal with it further here. Suffice it to say that this theory of dispersion stands in very good agreement with the results of experiment.[12]

An important optical discovery made in the year 1896 by the Dutch physicist, **Zeeman**, could also be explained on the basis of the theory of Lorentz. This is the **resolution of spectral lines in a magnetic field.** The so-called **Zeeman effect** is in general very complicated. In the simplest case, a spectral line is split into two lines in a magnetic field, these two lines being situated at equal distances on opposite sides of the position of the original line.[13] This

[11] We have an example of the production of forced vibrations in the fact that a bridge can be made to swing violently by suitably timed marching, because the periodicity of the marching acts in conjunction with the periodicity of the natural vibrations of the bridge.

[12] A very strong confirmation of the electron theory of dispersion is to be found in the fact that from an optical point of view all substances behave like metals towards waves whose frequency closely agrees with a natural frequency of an electron, i.e. such waves are very markedly reflected and absorbed. This phenomenon is called selective reflection and absorption. It also follows from the electron theory of dispersion that the idea of the dielectric constant must be more exactly defined than had been the case hitherto. Here we also find the explanation for the apparent deviations from the Maxwell relation (see note 16 of the first lecture).

[13] In the Zeeman effect we differentiate between the longitudinal and the transverse effect. We observe the longitudinal effect when

remarkable result could be deduced purely by mathematics from the electron theory developed by Lorentz, so that the Zeeman effect constituted a brilliant experimental confirmation of Lorentz' theory.

It followed from the more detailed investigation of the Zeeman effect that it must be called forth by **negative** electrical particles. Furthermore, measurement showed that the specific charge (i.e. charge per unit mass) of the electrical particles that give rise to the effect is about 1800 times greater than for ionized atoms of hydrogen. If it were assumed that the electrical particles are each charged with an elementary quantum of electricity, then we were forced to the surprising conclusion that the **mass** of the particles is still **1800 times smaller than** that of the **atom of hydrogen**, which had previously been held to be the smallest mass possible.

Within a few years this remarkable conclusion was completely confirmed by the results of the more exact investigation of the so-called **cathode rays.** The cathode rays, which were discovered in 1859 by **Plücker**, occur in Geissler tubes containing gases at low pressure, and they emanate from the place where the electric current leaves the tube,[14] i.e. from the cathode. The cathode rays travel in straight lines, and an object placed in their path throws a shadow in consequence; but their most remarkable property is that they are **easily deflected**

we are looking in the direction of the magnetic lines of force, and the transverse effect when we look in the direction perpendicular to the lines of force. In the simplest case, the spectral lines are resolved into doublets in the longitudinal effect, and for the transverse effect into so-called triplets.

[14] Cf. note 2.

by a magnet. Later it was found that they also suffer deflection in an electric field, and lastly, it was noticed that bodies become negatively electrified when cathode rays impinge upon them.

These properties of cathode rays can be very simply explained on the assumption that they consist of quickly moving electrically charged particles, which at the same time possess inertial mass. As a matter of fact, the rectilinear propagation follows immediately from the inertia of the particles, and the deflection in an electric field follows straightway from their electrical charge. On the other hand, since the swiftly moving electrical particles represent convection currents, they must suffer deflection in a magnetic field, exactly as a movable coil conveying a current is attracted or repelled by a magnet.

By the measurements of the deflections in an electric and in a magnetic field it was possible to determine two magnitudes of importance in connection with cathode rays; these were on the one hand the **velocity**, and on the other the **specific charge** of the cathode ray particles, i.e. the ratio between their charge and their mass. The results were very remarkable. It was shown that the velocity of the cathode rays so determined is exceedingly large, and that it amounts to from a fifth to a third of the velocity of light, which, as will be remembered, is equal to 300,000 kilometres per second. The cathode ray particles thus travel so quickly that they would only require about half an hour to complete the journey from the earth to the sun. The value obtained for the specific charge of the cathode ray particles was almost the same as for the hypothetical particles that were regarded on the electron theory as being the cause of the Zeeman effect. From this it was necessary to conclude that the particles responsible for the Zeeman effect are **identical** with the cathode ray particles, the only difference being that these **constituents of the atom** are bound to the atoms in the one case, whereas in the other they have been separated from the atoms and travel through space in the free state.

The so-called **positive rays** are likewise produced in Geissler tubes at low pressure, and constitute a counterpart to the cathode rays.[15] Like cathode rays, they are deflected in electric and magnetic fields, but in the reverse sense. Their velocity and specific charge could be determined by the same method as in the case of cathode rays. It was found that positive rays move more slowly than cathode rays, their velocity being only fractions of 1 per cent. of the velocity of light. However, the values obtained for the specific charge of the positive rays were of the same order of magnitude as for electrolytic ions. The mass of the positive ray particles must therefore be assumed to be almost as large as the mass of the atoms themselves.

The previously discussed phenomena connected with cathode rays offer a convincing proof of the correctness of the assumption, that an electron in motion gives rise to the same phenomena as a galvanic current. But since every circular galvanic current (as is known from experimental physics) produces the same effects as a magnet,[16] it follows

[15] The most important positive rays are the canal rays, discovered by Goldstein in the year 1886. They leave the perforated cathode in the opposite direction to the cathode rays when it is situated in the middle of the Geissler tube.

[16] We may recall the well-known properties of a solenoid bearing a current.

that the electron theory suggests the possibility of referring all the phenomena of **magnetism** to **revolving electrons.** Modern theory assumes that the molecules of all substances contain revolving electrons, which would thus represent **convective molecular currents.**[17] By means of this assumption all the previously known magnetic phenomena could be readily interpreted, particularly the fact discovered by **Faraday** in 1845, that **magnetism** is **a universal property** of all matter, and not solely of a few favoured substances.[18] On the other hand, the

[17] The idea of molecular currents can be traced to Ampère, who, although he did not regard them as convection currents, first endeavoured to explain magnetism by their aid.

[18] Experimental physics divides bodies into three classes according to their magnetic behaviour. Those substances that are repelled by a magnet are called diamagnetic; paramagnetic substances are those that are attracted by a magnet, and a group of substances that are very strongly paramagnetic are classed as ferromagnetic. To the last group belong in particular iron (whence the name), nickel, cobalt, and certain alloys discovered by Heusler in 1903, which are made up of metals themselves non-magnetic, e.g. an alloy of copper, manganese and aluminium. These three kinds of magnetism are explained on the electron theory by the assumption that a number of electrons are revolving in the molecules of all substances in such a way that they entirely or nearly compensate each other in their magnetic effects, in consequence of the differences in the sense of their revolutions. In the case of diamagnetic substances this compensation is regarded as complete, as long as there is no external magnetic field present. In the presence of a magnetic field (as in the approach of a magnet) induction currents result, which are superimposed on the molecular currents, and the direction of these currents is independent of that of the molecular currents upon which they are superposed. In virtue of this the compensation is incomplete, and the molecule itself thus behaves under the influence of the external magnetic field like a small magnet. From Lenz' law it follows that the sense of the induction currents must be such that the small magnet thus produced will be repelled by the large magnet. In the case of paramagnetic substances the com-

electron theory of magnetism also led to the discovery of new phenomena still unknown to the older theories.[19] But the most interesting result to which the electron theory of magnetism led was the recognition of a kind of parallelism that exists between the magnetic and the mechanical aspects of molecular currents.

A revolving electron possessing inertial mass represents a tiny magnet on the one hand, and on the other hand it is comparable with a **gyroscope,** from the mechanical view-point. But if this interpretation is true, then, as first recognized by **Einstein,** it must be possible by quick **reversal of its magnetism,** to cause a soft iron rod momentarily to **turn** by a slight though detectable amount. This experiment was successfully performed in the year 1915 by Einstein in conjunction with the Dutch physicist

pensation is regarded as being inherently incomplete, but these small magnets are nevertheless arranged indiscriminately, so that they give rise to no appreciable external effects. The conditions are altered, however, under the influence of an external magnetic field, as this arranges the small magnets. If the paramagnetism is stronger than the diamagnetism, which is a property of all matter, then the substance will not be repelled by a magnet, but attracted. Finally, ferromagnetism is explained on the electron theory by the assumption of special molecular forces.

[19] Amongst these we have primarily the fact discovered by Pierre Curie, that iron, nickel, and cobalt lose their ferromagnetism at definite temperatures, these being different for the different metals. This so-called Curie point lies, for instance, at 750° C. for iron, and at 375° C. for nickel. The theory explains this fact by supposing that at the Curie point the molecular forces responsible for ferromagnetism suddenly alter. It also follows from this assumption that a sudden alteration of the specific heat must occur at the Curie point, and that the magnitude of this jump must be connected with the magnetic properties. This has, as a matter of fact, been proved by experimental investigations.

de Haas. It followed from the observations that magnetic phenomena must evidently be produced by negative electrons. The reverse effect of that discovered by Einstein has also been observed, in that it has been possible to magnetize a soft iron rod by rotating it rapidly about its axis.[20]

Negative electrons, i.e. particles whose mass is only a tiny fraction of the mass of an atom, have thus been recognized as being really responsible for the production of the most varied phenomena. On the other hand, the properties of the positive rays show that **positively electrified particles** also exist, the mass of which is of the same order of magnitude as that of the atom. In point of fact, such positive and negative charges represent the constituents of which the atoms of all elements are composed. This became evident as the result of a group of phenomena that has attained such importance in physics and chemistry as to constitute at the present time a science in itself. We refer to the phenomena of **radioactivity.**

A few months after the discovery of the Röntgen rays, **Becquerel** made the surprising observation that compounds of the element **uranium** constantly emit **rays** without the intervention of any outside agency. These rays are able to affect a photographic plate through an opaque covering, and also to render the air conducting and thus to discharge a charged electroscope set up in the immediate neighbourhood. Substances such as the compounds of uranium and thorium, which emit these rays, are called **radioactive** substances.[21] After long and tedious investigations

[20] This effect was detected by the American physicist Barnett.

[21] *Radius* means—the ray.

Pierre Curie and his wife, **Marya**, discovered in 1898 that in pitchblende, from which uranium is obtained, there is contained in very small quantity an element which had up to that time been unknown. The radioactivity of this new element is millions of times more intense than that of uranium itself. M. and Mme. Curie called the new element **radium.**[22]

The years following the discovery of radium revealed clearly the nature of the **rays** emitted by radioactive substances. It was shown that these rays are of **three different kinds,** the α-, the β-, and the γ-rays. The β-rays are similar to the previously discussed cathode rays, the α-rays to positive rays, and the γ-rays to Röntgen rays. The **Gamma-rays** are, as mentioned in the first lecture, **Röntgen rays** of particularly short wave length; their frequency reaches values up to several hundreds of trillions per second.

In contrast with the γ-rays, which are a kind of light, both the α- and the β-rays consist of electrically charged particles possessing inertial mass. The proof of the correctness of this conception is supplied by the fact that both kinds of rays are deflected by a magnet. **Beta-rays** consist of negatively charged particles, and the value obtained for their specific charge is the same as for cathode rays. Hence β-rays evidently consist of **negative electrons.** The velocity of β-rays is still greater than that of cathode rays, and amounts to between 30 and 99 per cent. of the velocity of light. Measurements on the **Alpha-rays,** which consist of **positively** charged particles, have shown that their velocity lies between 5 and 7 per cent. of the velocity of

[22] The preparation of metallic radium was not accomplished until the year 1910 by Mme. Curie.

light. They thus move more slowly than the β-rays, but nevertheless more quickly than the positive rays to which they are essentially similar. The nature of the α-rays has been revealed by observations of a remarkable phenomenon known as **scintillation.**[23]

If we bring a radioactive preparation into the neighbourhood of a screen, the surface of which has been covered with zinc blende, we observe a continual flashing of discrete points of light. This suggests that each flash of light is due to the incidence of an α-particle. By observing a small area of the screen under the microscope, it was possible to **count** directly the number of **Alpha-particles** emitted in a definite time by an active preparation.[24] On the other hand, the total charge conveyed by the rays could also be determined.[25] Whence it became possible to evaluate directly the **charge of a single Alpha-particle**; the value thus found agreed accurately with that of **two positive elementary electrical quanta.** Moreover, from observations on the electrical and magnetic deflection of α-rays, their specific

[23] *Scintilla* means—the spark.

[24] In front of the radioactive preparation is introduced a screen made of some substance that does not transmit α-rays and in which there is a small hole. The opening is chosen so small that only those α-rays can pass through it which, when they strike the scintillation screen, can be observed in the microscope. The fraction of the total α-radiation passing through this opening is then calculated. For this purpose we only require to know the size of the hole and its distance from the preparation. By counting the number of flashes of light observed in the microscope in a definite time, the total number of α-particles emitted by the preparation in the time involved can be readily arrived at.

[25] By cutting out the β-rays emitted simultaneously with the α-rays by means of a magnetic field, a metal plate can be charged positively by the radioactive radiation.

charge is also ascertainable, i.e. the ratio between their charge and their mass. Thus when the magnitude of the charge had been determined, the value of the mass also became known. The result of this work was the surprising fact that the α-particles possess the same mass as the atoms of the gas helium,[26] the atomic weight of which is four, and which, after hydrogen, is the lightest chemical element. The α-particles are thus evidently none other than **helium atoms**, with the exception that they are not, like these, electrically neutral, but charged with two positive elementary quanta.

Apart from the emission of the three kinds of rays, the most remarkable phenomenon observed was the **continuous** and very large **production of heat** by radium, and this was at first quite inexplicable. It is so large that a quantity of radium would be able to raise the temperature of an equal mass of water from freezing-point to boiling-point once every three quarters of an hour. Physicists were intensely interested in the question of the origin of the relatively enormous amounts of **energy** that appeared in the radiation and in the heat production. This interest was intensified by the remarkable circumstance that radioactive radiation is in no way influenced by outer conditions such as that of temperature variation in particular, and, furthermore, that it is also completely independent of the chemical combination of the radioactive metal.[27] It was also a striking fact that radioactivity was observed just in the case

[26] The gas helium was first discovered in the atmosphere of the sun (hence the name ; *helios* = sun), but also later in the earth's atmosphere.

[27] As far as the radioactive radiation is concerned, it is immaterial whether the preparation is radium chloride or radium bromide.

of elements with the highest atomic weights (e.g. radium, thorium, uranium).[28] It was, however, the important discovery of the **emanations** that first indicated the real key to the solution of all these riddles.

Very soon after the discovery of radium, M. and Mme. Curie observed that bodies which were situated in the same vessel as radium themselves became radioactive. This phenomenon, which was at first called by the no longer suitable name of induced radioactivity, was then examined more closely by **Rutherford** in the case of **thorium.** In 1900 he established the surprising fact that the activity induced by thorium is brought about by the presence of a **gas** which is itself radioactive, this gas being continuously developed from thorium. Rutherford called this gas **thorium emanation.** In addition, Rutherford found that when the emanation is separated from the thorium it disappears, and that a radioactive **deposit** is formed from it, which, in consequence of its exceedingly fine distribution, coats with an invisible covering the surfaces of all bodies that have stood in contact with the emanation. Similar phenomena to those with thorium were found in the case of radium, the emanation of which was discovered soon after that of thorium. It was shown that, when **radium emanation** decays, other radioactive substances (solids) are successively formed, the production of each of which is accompanied by the simultaneous emission of rays. These substances are formed in such infinitesimally small quantities that they can only be measured by such

[28] The atomic weights of uranium, thorium, and radium are respectively (relative to oxygen = 16) 238, 232, and 226.

methods as the loss of charge experienced by a neighbouring charged electroscope, in consequence of the radiation. Nevertheless, the aggregate condition of all these substances could be recognized, their melting or boiling-points determined, and a judgment could also be formed as to their solubility in different acids.

On the basis of all these observations, **Rutherford and Soddy** in 1902 formulated a **theory of radioactivity** which, although excelling in its great simplicity, nevertheless signified a complete rupture with the hitherto deeply rooted and fundamental conceptions of physics and chemistry. The theory of Rutherford and Soddy, which is known as the **disintegration theory**, rests on the assumption of a **transformation of the atoms** of radioactive substances. In virtue of this, the science of radioactivity appears to be in a certain sense a completely new field of natural science. In the processes which physics had thus far been called upon to describe, the molecules remain unaltered. On the other hand, we call those processes chemical ones, which take place within the molecules without giving rise to alterations in the atoms. In contrast to this, however, the phenomena of radioactivity are dependent on processes which, according to Rutherford's theory, alter the **atoms** themselves.

The theory of Rutherford and Soddy regards the atoms as composed of positive and negative electrical particles, the number and geometrical arrangement of which determine the character of the chemical element under consideration. In spite of the minuteness of these charges, relatively strong forces of attraction and repulsion must act between them, in consequence of the very small distances involved.[29] A large **internal energy** must therefore be inherent in the atom. If the **condition of equilibrium** or configuration of the electrical particles in the atom is **not completely stable**, it can happen that the electrical particles **regroup** themselves into a new condition of equilibrium. But this would imply the **formation of a new chemical element**, which in its turn can be transformed into another element, and the more unstable the atoms are, the more rapidly will this transformation occur.

In the process of rearrangement of the electrical particles in the atom negative or positive particles or both kinds can be **ejected**, and this must take place with tremendous velocity owing to the relatively enormous internal energy of the atom. Thus we may explain the emission of the β- and α-rays. Further, the emission of the γ-rays is explained by the electromagnetic impulses that arise in the emission of the other two kinds of rays. Now as already mentioned, every α-particle has a mass equal to that of four hydrogen atoms. Hence the new element formed by an α-ray radioactive transformation must have an **atomic weight less by four** (or 8, or 12) **units** than that of the so-called parent substance, and conversely. As a matter of fact, the atomic weight of radium emanation (222) was found to be 4 units less than that of radium, and the atomic weight of radium 12 units less than that of uranium (238), from which it is formed by α-transformation after the intermediate production of two other α-rayers.

Rutherford showed directly by means of the spectroscope that, as their atomic weight suggests,

[29] According to Coulomb's law, the forces are inversely proportional to the square of the distance.

the particles of the α-rays are in reality identical with atoms of the earlier known gas **helium**, after their charge has been neutralized. **Ramsay** and **Soddy** had previously proved the **production of helium during the decay of radium emanation**, and were thus the first to discover that an already known element can be produced from another one.[30]

The probability of the chance **disintegration** of an atom within a definite time depends on the **stability** of the arrangement of the electrical particles in the atom. The more unstable the arrangement, the greater is this probability. Since in general the arrangement is likely to be more unstable the greater the number of the electrical particles combined in the atom, we can grasp why it is that the phenomena of radioactivity can be most readily detected in the case of elements of high atomic weight.

The time required in order that an arbitrary quantity of a substance may be reduced to half in consequence of the disintegration of its atoms is called the **half-value period** of that substance. In the case of strongly radioactive substances the half-value period can be determined by observation of the decrease in the intensity of the radiation, and for radium it can also be calculated by enumeration of the α-particles emitted. The half-value period of radium is 1580 years, of radium emanation about four days, and of thorium emanation only about one minute. The half-value period of substances of slight activity can be calculated on the basis of a relation that can be derived theoretically; viz. the

[30] We are also led to regard radium emanation, for instance, as an element, from the fact that it possesses a characteristic spectrum.

quantities of the parent substance and of the disintegration product contained in a preparation must bear the same ratio to each other as the half-value periods of the two substances. For instance, it has been found that in uranium minerals one gram of uranium is associated with an amount of radium equal to about a three-millionth part of a gram. The half-value period of **uranium** must therefore be about three million times as large as that of radium, i.e. it must amount to about five milliards of years. If we had a billion atoms of uranium we should therefore find that on the average one would disintegrate about every three days.

Although radioactive radiation can readily be detected from the atoms of uranium, their stability is nevertheless exceedingly great. In the case of an element which does not show radioactive phenomena, it follows that the probability that an atom of this element will disintegrate even in a very large interval of time must be excessively small and still less than for an atom of uranium. At the same time we can hardly believe that from amongst the elements just those known to us as radioactive should occupy a unique position, and differ from all the others in their nature. It does not seem unlikely, then, that in **gradual disintegration** we perceive a **universal property of all matter**, and that it occurs only in different degrees for different elements.

Our **conception of the chemical element** has thus experienced a complete **transformation** as a result of the discovery of radioactive processes. The inviolable dogma of the nineteenth century of the absolute, rigid immutability of the elements has fallen; a transformation of the elements is not only possible, but it has been indubitably established. Like organic existence, the elements now appear to us to be transitory and of limited life. But above all, the new physics has led to the conclusion that the numerous elements between which chemistry differentiates are nevertheless of an entirely **similar** nature. They are all built up of positive and negative electrical particles, and only the number and arrangement of these differ. From the fact that both the Zeeman effect and also the phenomena of magnetism are produced by negative electrons, it was necessary to conclude that within the atom negative electrons, which only constitute a very small part of the atomic mass, are in rapid periodic motion. On the other hand, the total positive charge of the atom, if it appears **neutral** as a whole, must be just as large as the sum of the charges of all the negative electrons contained in the atom. The atomic theory was thus confronted with the fundamental problem as to how we have to regard the **positive charge** as being distributed **within the atom.** Information on this question was obtained from the remarkable results of the experimental research of the English physicist, **C. T. R. Wilson.**

A more detailed investigation of the α-rays showed that each single α-particle, by virtue of its electrical charge and its velocity, can produce **condensation nuclei** for water drops in the gas through which it passes.[31] By means of a suitable arrangement of apparatus Wilson thus succeeded in making the paths of single α-particles appear as fine fog-streaks. A photograph of these supplies us with

[31] It had already been known for some time that the ions of the air constitute condensation nuclei for the formation of clouds.

a **picture of the** actual **paths of the α-particles.** In general the photographs show **straight lines,** some of which, however, appear **bent** through fairly large angles. It is therefore very probable that these large deflections are due to a single collision between an α-particle and an atom. On the other hand, α-rays penetrate through a distance of several centimetres of ordinary air, whence it follows that an α-particle passes through many thousands of atoms. On the basis of Wilson's photographs it was thus necessary to assume that an α-particle can penetrate through many thousands of atoms without its direction being appreciably affected, whilst occasionally a single atom gives rise to a deflection through a very large angle. This remarkable fact can be simply explained on the assumption that the deflection is caused by a force of repulsion exerted on the positive α-particle by the positive charge of the atom penetrated, and that this positive charge is concentrated in a volume which constitutes only a very small part of the total volume of the atom. Wilson's experiments constitute an important empirical confirmation of a **theory of the atom** that had been previously developed by **Rutherford,** according to which atoms contain a **positively charged nucleus,** round which negative electrons revolve like planets round the sun. Nevertheless, the further development of Rutherford's theory only became possible on the basis of the quantum principle, which we shall discuss in the next lecture.

As a result of the electron theory, a conception in natural philosophy just as old as it is fundamental has completely altered the significance attached to it during many hundreds of years. The **idea of**

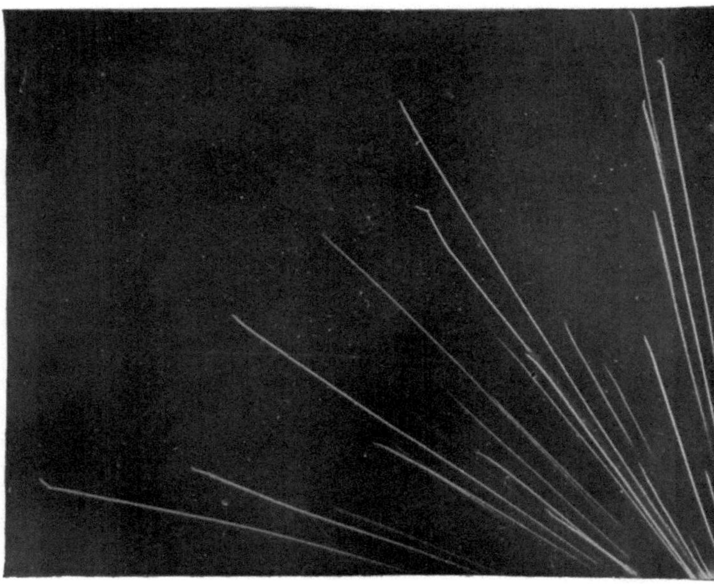

FIG. 2

[*To face page* 70

matter in its original sense no longer exists for modern physics. Until recently, physics dealt with electricity in its relation to massive matter, and regarded electricity as nothing else than a state of massive matter. Later research has shown that the massiveness of matter is only a consequence of its electrical constitution. In the modern system of physics electricity no longer stands alongside of matter; it has taken the place of matter. The new physics can descry in **electricity** that **unadulterated primordial something** for which scientists sought through thcusands of years,[32] and from which all things amenable to sense-perception are formed.

[32] The problem of the primordial substance, which Thales founded, constituted the starting point of Hellenic philosophy.

IV. THE QUANTUM THEORY

IN the kinetic theory of heat and in the electron theory the **atomistic principle** had already shown its exceeding fruitfulness as a result of its application to the internal state of motion of matter and to electricity. The **quantum theory** arose in an endeavour to extend still further and more generally this well-established principle. Originating in 1900 from a problem in the theory of thermal radiation, the quantum theory has since then made rapid strides, and is far from having reached the end of a development that has led physicists from one success to another. It not only made possible the formulation of a general law of radiation, but also led to new ideas on the progagation of light. It explained the behaviour of bodies at very low temperatures, which had been unintelligible to classical physics. But the quantum theory scored its greatest triumphs in a domain that first gained a place in the system of theoretical physics by virtue of the theory itself. We refer to the study of those spectral phenomena through which the internal structure of the atoms has been revealed.

The kinetic theory of heat and the electron theory had applied the atomistic principle to the **objects** of physical happening; the quantum theory applies the atomistic principle to the physical processes themselves. Just as physics had previously used an elementary quantum of mass and one of electricity as the basis of its considerations, so the quantum theory **reduces to elementary units** a physical magnitude, which in a certain sense measures the physical processes themselves. The great importance of this magnitude, the so-called **action** of a process, had been recognized by physicists even in the eighteenth century. It is obtained by multiplying energy magnitudes by time magnitudes. The quantum theory, then, is based on the assumption that the action of physical processes is built up from **elementary quanta of action.**[1]

This idea was first applied to the **theory of thermal radiation.** From the modern point of view the theory of radiation is regarded as the study of the **inter-relations** and interactions between the **electromagnetic waves** filling space and the **micro-mechanical processes** known to experimental physics as heat. The newer theory of radiation is based on an important law formulated by **Kirchhoff** in the year 1859, which has reference to the **emissive power** of a body. By this we understand the amount of energy radiated per second from a warm body, per 1 sq. cm. of the surface of the body. Kirchhoff was led to the conclusion that, apart from the temperature, the emissive power of a body depends solely on the degree to which the body **absorbs** heat radiation falling upon it. If it absorbs the heat radiation, or, as we must now say, if it absorbs the incident electromagnetic waves completely, so that none of the waves are reflected, we then call the body

[1] Mechanical processes are determined by the fact that for them (for a particular kind of variation of the motion) the action represents a minimum; but the action is determined by the time integral of the difference between kinetic and potential energy.

perfectly **black.**[2] Hence, according to Kirchhoff, the **emissive power of a black body depends simply and solely on the temperature.**[3]

The nature of this dependency was first discovered in 1879 by **Stefan,** but it was **Boltzmann** who first gave an exact theoretical foundation for the law found by Stefan.[4] According to the so-called Stefan-Boltzmann law, the **emissive power** of a body is **proportional to the fourth power of the absolute temperature** of the body. The ratio between the emissive power of a black body and the fourth power of its absolute temperature must therefore represent a **universal constant,** which is called **Stefan's constant.** Its value can be determined by observing the cooling of a hot body.

The radiation emitted by a warm body, however, is composed of most varied periods, of most different wave lengths. The totality of all the possible periods and thus of all the possible wave lengths from the shortest to the longest is called the **spectrum,** in the broadest sense of the word. It is a matter of everyday observation, however, that all the parts of the spectrum by no means occur in equal intensity

[2] The opposite of perfectly black is perfectly reflecting.

[3] The existence of so-called thermodynamic equilibrium is nevertheless a necessary assumption for the validity of Kirchhoff's law. In thermodynamic equilibrium all bodies belonging to the system considered have the same temperature or at least they have not appreciably different temperatures.

[4] In this derivation radiation pressure plays an important part. From Maxwell's theory it follows, and Maxwell himself recognized this, that a body upon which electromagnetic waves are incident must in consequence experience a pressure, which may be termed light pressure, or in more general terms, radiation pressure. The existence of light pressure was first proved experimentally by Lebedew in 1901, with the aid of very light mirrors in a very highly evacuated vessel.

in the radiation of a warm body, and that the share of each of these in the total radiation varies with the temperature. When a body is heated, it at first emits only dark heat rays, because the part of the spectrum visible to the eye is only inappreciably represented in the radiation. Not until 525° C. does the body begin to glow visibly,[5] first with a reddish glow, and then as the temperature is still further increased we get a yellowish glow, and finally the body becomes white hot.

Theoretical physics was thus confronted with an important and at the same time exceedingly difficult problem as to the **spectral distribution of the energy of radiation.** The idea involved may be illustrated by an example taken from a domain quite apart from physics. In economics, statistical investigations are made of the distribution of the people's income in the different grades. It constructs **grades** of annual income which increase by a constant amount of, say, five pounds, and asks the question as to the percentages of the total annual people's income that fall into the different grades. We should find that one of the grades was unique in that its percentage share of the total people's income was greater than any of the others. In quite a similar way theoretical physics can grade the spectrum according to wave length,[6] and investigate the fundamental question as to the distribution of the total energy radiated by a body amongst the individual grades of the spectrum at

[5] The temperature at which the visible glow sets in is, as was first recognized by Draper, the same for all bodies ; this again is a consequence of Kirchhoff's law.

[6] The distribution of energy over the grades (steps) of the spectrum is represented by other formulæ, when the grading is performed not according to wave length but according to frequency.

particular temperatures, or, in other words, their individual shares of the total emissive power. Here, too, we should find that there is a definite part of the spectrum for which this share is the greatest. If the intervals (grades) are chosen very small, a definite **wave length** can thus be found for which the so-called **specific emissive power** is **greatest**.

In the year 1893 **Wien** made the important theoretical discovery that the product of this wave length and the absolute temperature of the radiating body constitutes a **universal** constant, thence called **Wien's constant.** Thus the higher the temperature, the smaller is the corresponding unique wave length. The position in the spectrum corresponding to this wave length is thus displaced with increasing temperature in the direction from larger to smaller wave lengths, or within the visible spectrum in the direction from red to violet.[7] The law found by Wien was therefore called the **displacement law.** The correctness of this law has been completely confirmed by experiment, and in this way it was also possible, on the other hand, to determine Wien's constant with moderate accuracy.[8]

The discovery of the displacement law constituted a great and fundamental success in the task of solving the basal problem of the radiation theory, namely,

[7] For the sun, a temperature of 6000° C. would result from Wien's law, were we able to regard the sun as a black body. On the same assumption we can also calculate the sun's temperature from Stefan's law, since we can measure the heat received in a given time by a black surface of given area, on which the sun's rays fall. From this we can find the total amount of energy radiated per second by the sun, and thence calculate the sun's temperature by Stefan's law. In this way we also arrive at a temperature of about 6000° C.

[8] The value of Wien's constant is 0·294 cm. degrees.

the question as to the spectral distribution of energy in its dependency on the temperature. This problem was too much for classical physics. In two quite different ways it had certainly led to two different laws that were to represent solutions of the problem, but neither of the two laws was able to stand the test of experiment. It was found that they both are valid only within a limited range, and that neither of them possesses general validity.[9] The complete solution of the radiation problem was not arrived at until 1900, when **Planck** devised the ingenious hypothesis of the **elementary quantum of action,** and applied it to the phenomena of radiation.

Planck made the assumption that the emission of radiation occurs discontinuously, in such a way that **elements of energy** play a part in the process. Moreover, the magnitude of these was to be determined by the condition that the product of an element of energy and the time of its oscillation period should be equal to the elementary quantum of action. Since the frequency is the reciprocal of the oscillation period, it follows that each element of energy must be equal to the product of the elementary quantum of action and the frequency. Thus the energy elements of differently coloured radiations will not be equally large, but much smaller for the infra-red than for the ultra-violet. For violet light they would be about twice, for green light about one and a half times as large as for red light.

By means of the hypothesis of elements of energy,

[9] One of these laws was derived by Wien from his displacement law; it is valid only for the ultra-violet spectrum or for low temperatures. The other law was established by Rayleigh; it is valid only for the infra-red spectrum or for high temperatures.

Planck was able to derive **a law of radiation** that correctly represents the distribution of the energy of radiation over **all parts of the spectrum at all temperatures**, which is in **good agreement with experimental observation.**[10] But the importance of Planck's law of radiation reaches far beyond the domain of the actual study of radiation, for it led directly to the accurate determination of the most important universal constants of physics. Two equations follow from Planck's law, and these link up the empirical constants of Stefan's and of Wien's laws with two fundamental magnitudes in the modern system of physics. One of these magnitudes was quite unknown, and the other had been only roughly estimated. The former was the hypothetical elementary quantum of action, and the latter was the **mass of the hydrogen atom**, for which up to that time only the order of magnitude had been estimated by Loschmidt. By solving the equations referred to above for the two unknowns, Planck found very accurate values for the elementary quantum of action and for the mass of the hydrogen atom. For the latter he obtained the value of a quadrillionth part of 1½ grms.,[11] already given in a previous lecture, and from this Planck was further able to calculate very accurately the elementary quantum of electricity, by making use of known electrochemical constants.

[10] The difficult experimental investigations, requiring great accuracy, which proved the non-validity of the earlier and the validity of Planck's law of radiation, were carried out in the Physikalisch-Technischen Reichsanstalt in Berlin-Charlottenburg, mainly by Rubens and Paschen.

[11] It should be noted that the mass of an atom is only half as big as that of a molecule.

The value that Planck obtained for the **elementary quantum of action** appears to be very small from the human point of view. We can form an idea of its magnitude when we consider energy elements of visible radiation. Although the frequency of violet radiation reaches the tremendous value of 800 billions per second, the energy elements would nevertheless be so small, that it would take 2 trillions of them to make up the energy that would have to be expended to raise a weight of 1 kg. to a height of 1 metre.[12]

In 1905, five years after Planck had so successfully founded the quantum theory, **Einstein** discovered a new and important field of application for the quantum principle in a group of phenomena, in which light is transformed into light of another period, or energy of motion into light, or finally the converse process in which light is transformed into energy of motion. The first process—the transformation of light into that of another frequency—occurs in the phenomenon of **fluorescence**, which is observed both with visible light and also with Röntgen rays.[13] The second process, in which light is produced by energy of motion, is observed when cathode rays, which have inertial mass and a relatively enormous energy of motion, give rise to Röntgen rays. The third process, the production of energy of motion by light, is known as the

[12] The value of the elementary quantum of action, generally denoted by h in modern physics, is $6 \cdot 5 . 10^{-27}$ erg-seconds; cf. note 7 in the third lecture. An erg is the work done when a force of 1 dyne acts through a distance of 1 cm.; about 98 millions of ergs are equal to 1 kilogram-metre.

[13] For the phenomenon of fluorescence the law of Stokes holds, according to which the frequency of the fluorescent light is always smaller than that of the exciting light.

photoelectric effect. In this, electrons are torn from the bodies subject to radiation by ultra-violet light or by Röntgen rays.

Einstein found a simple interpretation for remarkable facts which had attracted the attention of research workers on the phenomena mentioned. He assumed that the elements of energy play a part not only in the emission of radiation by warm bodies, but also that light itself is **propagated in light quanta,** the magnitude of which is determined by the product of the elementary quantum of action and the frequency. On the basis of this hypothesis, Einstein derived a law for the photoelectric effect that was completely substantiated by experimental observation.[14] In fact, Einstein's law is so exactly obeyed, that the American physicist, **Millikan,** was able to make use of a method for the direct determination of the elementary quantum of action, which was founded on the measurement of the photoelectric effect. The value so obtained by Millikan completely agrees with that found by Planck by quite a different method.

In 1907, two years after he had founded the idea of light quanta, **Einstein** discovered an additional very important field of application of the quantum hypothesis, in the theory of the **specific heat of solid bodies.** By the specific heat of a substance we understand the amount of heat that must be supplied to 1 grm. of the substance to raise its temperature by 1° C. In 1819 **Dulong** and **Petit** discovered an interesting relation between the specific heat of a solid element and its **atomic weight.** According to

[14] Einstein's law states that a light quantum of the incident light is equal to the energy of motion of an electron separated by the photoelectric effect plus the work that must be expended to free it.

the so-called law of Dulong and Petit, the product of these two quantities, the **atomic heat,** should have the **same value for all solid elements.**[15] This relation, a theoretical basis for which was given later, is in general well confirmed by experience. But even at the time of the formulation of the law, very large **deviations** from the law were noted in the case of several elements of low atomic weight, e.g. beryllium, boron, and particularly diamond. Later, the fact was established that especially in the case of diamond the specific heat decreases very much when the crystals are strongly cooled.[16]

The deviations from Dulong and Petit's law could not be explained by classical physics. But they certainly found a simple interpretation when **Einstein** applied the **quantum principle** to the **theory of heat of solid bodies.** In accordance with the classical ideas, Einstein assumed that the heat of solid bodies is based on the **vibrations** of their atoms, but he further assumed that the energy of these vibrations is made up of **elements of energy** in the sense of Planck's

[15] This value is 5·94 cals. Compare note 3 in the second lecture.

[16] At room temperature the atomic heat of diamond is only 1·7 cals.; at − 50° C., as was discovered as far back as 1875, it is only 0·7 cals. The value of the atomic heat of diamond at − 187° C. is only 0·03 cals., and at − 250° C. 0·00 cals.; i.e. at this low temperature the existence of a specific heat is no longer detectable, in spite of the measurements being accurate to the second decimal place. The diminution of the specific heat at low temperatures was established recently also in the case of other materials. With copper, for instance, which behaves normally at room temperature, the value of the atomic heat at − 186° C. is only 3·38, at − 240° C. 0·54, and at − 250° C. only 0·22 cals. The experimental work on the behaviour of bodies at the lowest attainable temperatures was carried out mainly by Kamerlingh Onnes in Leyden and by Nernst and his co-workers in Berlin.

hypothesis, according to which these elements are proportional to the frequency. As a result of this assumption of Einstein, not only did the **decrease of the specific heat** at low temperatures become intelligible, but also the fact that even at room temperature appreciable deviations from Dulong and Petit's law occur with elements, the atoms of which, being particularly light, vibrate very rapidly in consequence.[17] Einstein's idea was subsequently developed by different scientists, in particular by **Debye**, who improved the theory of Einstein and thus brought it into still better agreement with experience. In 1912, Debyé's work led him to an important law, according to which the **specific heat of a solid body is proportional to the third power of the absolute temperature** at very low temperatures.

The deductions from Debye's law agree with very important conclusions which **Nernst** had previously obtained from a principle that he formulated in the year 1906. This principle, generally known as **Nernst's heat theorem**, produced entirely new foundations for the theory of heat and the thermochemistry of low temperatures, and in many ways proved its extraordinary fruitfulness and its good agreement with the results of experience. The purport of Nernst's heat theorem is of course difficult to bring into a form intelligible to the layman; in its essence the law says that at very low tempera-

[17] For every solid element a so-called characteristic temperature can be determined theoretically. Only when the temperature at which the measurements are performed is somewhat higher than the characteristic temperature does Dulong's law have validity according to the new theory. Now for most solid elements this characteristic temperature lies below the melting-point of ice, whereas for diamond it lies at about 1700° C.

tures the idea of temperature loses the significance attached to it under normal conditions. At very low temperatures the specific heat decreases rapidly as the temperature falls. A quantity of heat which at room temperature would not be able to raise the temperature of a body appreciably can produce a rise of many degrees when the temperature is very low. A quantity of heat, which at room temperature appreciably expands a body, is unable to effect an observable enlargement of the volume of the body at very low temperatures.[18]

The universal character of the quantum hypothesis had thus already been recognized in three quite different directions—in the theory of thermal radiation, in the theory of the photoelectric effect, and in the theory of specific heat. At this juncture, in 1913, the Danish physicist **Bohr** opened up a new field to the application of the quantum theory, and here again it gained fresh laurels. By incorporating the hypothesis of the elementary quantum of action with **Rutherford's theory of the atom,** Bohr was enabled to develop a **theory of spectra,** which in its further development at the same time brought the solution of the difficult problem of the **structure of the atom.**

In the year 1859 **Kirchhoff** and **Bunsen** had made the important discovery that in the **spectra of the chemical elements lines of definite frequency** occur, which are characteristic of these elements. **Balmer** then discovered in 1885 the exceedingly noteworthy fact that **very simple numerical relations** exist **between the lines of the spectrum of hydrogen.** The frequencies of the individual hydrogen lines, both those

[18] Like the specific heat, the co-efficient of expansion also approaches the value zero at very low temperatures.

known to Balmer and those that were not discovered until later, can be represented by **differences** of the form $R/m^2 - R/n^2$, where both **m** and **n** are always simple **whole numbers.** Here **R** represents a constant frequency of 3291 billions a second, i.e. a frequency that lies about two octaves higher than that of visible violet light.[19]

In the nineties of the nineteenth century the Swedish physicist **Rydberg** had already made the important discovery that the oscillation number **R**, thence known as the **Rydberg constant**, not only determines the structure of the hydrogen spectrum, but that it also plays an important part in the more complicated formulæ describing the regularities of other chemical elements.[20] In this way the Rydberg frequency achieves the significance of **a universal** constant. In the year 1910 the author of this book, who was the first to apply the quantum principle to the theory of the atom and to the theory of spectra, found a relation that links up the Rydberg constant with the fundamental magnitudes of the quantum theory and the electron theory, i.e. with the elementary quantum of action, the elementary quantum of electricity, and the mass of the negative electron.[21]

In 1913 **Bohr** succeeded in obtaining a surprisingly simple interpretation of the regularity of the hydro-

[19] In no region of physical measurement is such a degree of accuracy attainable as in spectroscopy. Spectroscopic magnitudes can often be given to the millionth part of their value, sometimes still more accurately.

[20] These regularities were discovered particularly in the case of the alkali metals (lithium, sodium, potassium, rubidium, and caesium) by Kayser and Runge.

[21] See Fritz Reiche, "The Quantum Theory," p. 172, note 249. Translation by H. S. Hatfield and H. L. Brose. (Methuen, 1922.)

gen spectrum, by taking as the basis of the quantum theory of the atom the **atomic model** that **Rutherford** had devised shortly before. According to Rutherford's conception of the atom, which has already been discussed in the third lecture, the atom of hydrogen is regarded as being composed of an **atomic nucleus** consisting of a **positive electron** (**proton**),* and of a **negative electron** that **revolves round** the nucleus. The negative electron comprises only about the 1840th part of the mass of the atom, whereas the rest of the mass must be ascribed to the atomic nucleus.

Bohr's application of the **quantum theory** to Rutherford's model of the hydrogen atom was of a **dual** nature. His first assumption has reference to the circular motion of the negative electron. Whereas according to classical physics this motion can take place in orbits of quite arbitrary diameter, without any one orbit having preference over any other, Bohr assumed that those **orbits have precedence**, for which a definite magnitude [22] characteristic of the motion from the mechanical point of view is a **whole number multiple of the elementary quantum of action.** According to the magnitude of this whole number, we therefore speak of a **one-quantum orbit**, of a two-quantum orbit, of a three-quantum orbit, and so on. To every orbit of definite **quantum**

* See translator's note.

[22] This magnitude is 2π times the angular momentum, which is measured by the product of the mass, the radius of the orbit, and the linear velocity. If the angular momentum is given, then all the other magnitudes which play a part in the motion can be calculated from the relation which states that the electrical attraction calculated from Coulomb's law must be equal to the centrifugal force.

number there are quite definite corresponding values of the radius of the orbit, of the velocity, of the revolution number, and of the energy. For the **one-quantum orbit,** for instance, the velocity amounts to about a 140th of the velocity of light, and the revolution number is about 6000 billions a second. The energy possessed by the atom of hydrogen in the one-quantum state is simply equal to the product of the Rydberg constant and the elementary quantum of action. The value found for the radius of the one-quantum orbit is about the twentieth part of a millionth of a millimetre.

The values which hold for **many-quantum** states stand in an exceedingly simple relation with the values given by Bohr's theory for the one-quantum state of the atom. The radius corresponding to an orbit of arbitrary quantum number is found by multiplying that of the one-quantum orbit by the square of the quantum number. Thus the radius of the two-quantum orbit is four times, that of the three-quantum orbit nine times, and that of the four-quantum orbit sixteen times as large as the radius of the one-quantum orbit. The converse holds for the energy. In the two-quantum state the energy of the atom is only a quarter, in the three-quantum state only a ninth, and in the four-quantum state only a sixteenth part of the energy in the one-quantum state. We obtain the **energy** associated with one of these multiple-quantum privileged states by **dividing** the energy of the one-quantum state **by the square of the quantum number.**

From what has already been said, on the other hand, the frequency of a line in the hydrogen spectrum can be expressed as the difference of two terms, each of which contains the square of a whole number in the denominator, the numerators, however, being equal. In order to be able to explain the regularity of the hydrogen spectrum, Bohr thus only required to supplement his first assumption by a second, which results directly from the quantum principle and in particular from Einstein's hypothesis of **light quanta.** Bohr assumed that the atom only emits light when it passes over from one privileged or stationary state to another stationary state, and that the energy which becomes available in virtue of this transition is transmuted into a light quantum. In other words, the difference between the energy values corresponding to the two stationary states must be equal to the product of the elementary quantum of action and the frequency of the light emitted. As a result of this **frequency condition,** the frequency of the spectral line produced by the transition will be determinate. Since the values of the energy are inversely proportional to the squares of the quantum numbers, the frequency condition makes it clear that the frequencies of the hydrogen lines are representable as the **differences** of two fractions, each having the same numerator but containing in the denominator the squares of different whole numbers. On the other hand, the frequency condition appears to link Rydberg's constant with the magnitudes that characterize the hydrogen atom from the standpoint of the electron theory. Now on the basis of this relation it is also possible from spectroscopic measurements to calculate very accurately the elementary quantum of action; the value so found completely agrees with that derived by Planck from his radiation law, and also with that obtained later by Millikan from photoelectric measurements.

Since every spectral line is produced by a transition or jump between **two** states, there results as it were a **double manifold of spectral lines.** To every end state of definite quantum number there belongs a **series** of spectral lines, the single lines of which are determined by the quantum numbers of the initial states. (The number **m** determines the end state in the emission, and the number **n** the initial state; for a given **m, n** can pass through the series of whole numbers.)[23] The series visible to the human eye, the so-called **optical series (Balmer series)** of the spectrum of hydrogen is produced by a jump to the **two-quantum end state**; for the jump from the three-quantum to the two-quantum state a red line is produced, and for the jump from the four-quantum to the two-quantum state a blue line results.[24] As the quantum number of the initial orbit increases, the lines of the optical series given by the Balmer formula approach nearer and nearer to a frequency which is equal to the Rydberg constant divided by the square of 2. This limit of the optical series lies in the ultra-violet. Up to the present a total of thirty-three lines has been observed in the optical series.[25] The series corresponding to the one-quantum end state lies in the ultra-violet, and the

[23] Conversely the hydrogen atom only absorbs from the incident radiation waves of such frequencies as satisfy the Balmer formula. Here the negative electron passes over from an initial state determined by the number *m* to an end state determined by the number *n*. This is the reverse of the case of emission.

[24] The red line is identical with the Fraunhofer C-line, and the blue one with the Fraunhofer F-line.

[25] Only the first twenty lines of the emission spectrum have hitherto been observed in low pressure Geissler tubes; the remainder have only been found in the spectra of nebulæ. In the absorption spectrum as many as fifty lines can be absorbed in laboratory experiments.

series corresponding to a three-quantum end state is situated in the infra-red. Here we have an explanation of the fact that these two series were discovered later than the optical series,[26] since they are more difficult of detection.

Be it briefly mentioned that Bohr's theory also led to a simple interpretation of the **spectrum of helium**;[27] in fact it was due to his theory that series which had previously been **wrongly ascribed to hydrogen** were recognized as helium series.[28] It was a triumph for Bohr's theory when these line series, which had previously been erroneously referred to hydrogen, were actually observed later in Geissler tubes filled with pure helium and free from every trace of hydrogen, in accord with the prediction of theory.[29]

Bohr himself obtained a still more brilliant confirmation of the new conception of the spectrum of helium. This work, which was purely theoretical, was based on the remarkable fact that in the helium spectrum the agreement between the theoretically calculated and the actually observed frequencies of the lines did not appear so complete as in the case of hydrogen. As has often happened in the history of physics, it was found in the present instance that these supposed deviations from the theory

[26] The ultra-violet series was discovered by Lyman, and the infra-red by Paschen in the spectrum of a Geissler tube.

[27] Helium is so important in theoretical physics because, along with hydrogen, it is the lightest and thus presumably also the simplest of the elements.

[28] One of these series was discovered by the astronomer Pickering in the spectrum of a star, and two others by Fowler in the spectra of Geissler tubes. The different series of hydrogen and helium are generally called after their discoverers.

[29] This was demonstrated by Paschen in 1916.

nevertheless proved themselves to be a brilliant confirmation of the theory in its perfected state.[30] Bohr modified the theory in this case by taking account of the **motion of the atomic nucleus.** As a matter of fact, the negative electron does not revolve round the nucleus in Bohr's model of the hydrogen atom (to which the helium atomic model is analogous),[31] but both the negative electron and the nucleus revolve round their common **centre of gravity.**[32] When Bohr introduced this previously disregarded fact into his considerations, he not only completely succeeded in explaining the supposed deviations in the spectrum of helium, but was able accurately to calculate from the magnitude of these supposed deviations a fundamental constant of the electron theory, viz. the ratio of the charge to the mass of the nega-

[30] We recall how the perturbations of the orbits of planets were first regarded as showing the insufficiency of Newton's theory of gravitation, until Laplace, by an exact development of the theory of perturbations, supplied proof of the fact that the supposed deviations are in reality confirmations of Newton's law of gravitation.

[31] The theory regards the helium atom as consisting of a nucleus possessing two positive elementary quanta of electricity acting outwards and of two negative electrons that revolve round the nucleus. If the neutral helium atom parts with an electron, it will then appear to the outside world to be charged with one positive elementary quantum, i.e. "ionized" (cf. the third lecture). Thus in the case of an ionized helium atom, exactly as with a hydrogen atom, only one electron revolves round the nucleus, so that this electron is not disturbed by any other similarly revolving electrons. Hence the conditions are just as simple as in the case of the hydrogen atom. Nevertheless the charge of the nucleus is twice that of the hydrogen atom, and it is owing to this that, according to Bohr's theory, the Rydberg constant appears in the formulæ for the helium series multiplied by the square of 2.

[32] The common centre of gravity of two bodies lies along the line joining them, and divides this line in the inverse ratio of the masses.

tive electrons. The value obtained by Bohr in this way was in complete agreement with that already obtained in quite a different manner from measurements with the cathode rays, and constituted a brilliant triumph for his theory.

Two years after Bohr had so successfully evolved the quantum theory of spectra, i.e. in 1915, this theory was the subject of an important extension by **Sommerfeld.** The relation of this physicist to Bohr is comparable, as it were, with that of Kepler to Copernicus. Just as at one time Copernicus thought of the orbits of planets as being circular, whereas Kepler treated them as ellipses and thus wrought a tremendous advance, so Sommerfeld replaced the circular orbits of the Bohr atomic model by **elliptical** ones. The theory naturally became much more complicated in consequence of this, for a circle is determined by a single magnitude, its radius, whilst an ellipse requires two magnitudes to determine it, viz. its major and its minor axes. Thus on Sommerfeld's theory two quantum numbers are necessary for the identification of every permissible orbit from the standpoint of the quantum theory. Spectral lines thus represent a **quadruple manifold** in Sommerfeld's theory, since both the initial state and the end state require two quantum numbers to describe them.

When Sommerfeld set about to develop the theory of spectra on this basis, he at first obtained just the same lines as Bohr. The sole difference was, that whereas on Bohr's theory only one possibility of production of each line was known, there were several possibilities of producing each of these lines on Sommerfeld's theory. At this stage, then, the ellipse hypothesis appeared to be only a superfluous and useless complication of the earlier theory. However, the new hypothesis demonstrated its extraordinary usefulness when Sommerfeld combined it with a fundamental law of the **theory of relativity.** We refer to the law, discussed in the sixth lecture, according to which the **mass** of a body with respect to an observer **depends on the velocity** of the body relative to that observer. Since the velocity of the negative electron in the atom of hydrogen amounts to a few thousandths of the velocity of light, it is quite possible that appreciable deviations from classical mechanics occur.

By taking into consideration the dependence of mass on velocity, Sommerfeld found that for the several transitions (jumps) which on his theory take the place of a single Bohr jump, not exactly the same line arises, but different lines, although the frequencies of these amongst themselves differ only very slightly. Thus according to Sommerfeld's theory every Bohr line appears **resolved** into a **group** of lines exceedingly close to each other. The structure of this group is called the **fine structure** of what is in Bohr's theory a single line. Nevertheless, in the case of hydrogen, the fine structure cannot be completely recognized even with the best spectroscopic apparatus of the present time. But in spite of this, the fact was already known that in spectroscopic apparatus of high resolving power the individual lines of the optical series of hydrogen appear doubled, or, in other words, they form **doublets.** This corresponds to the case in which each group that occurs in the place of a line in the optical series of hydrogen on Sommerfeld's theory is further composed of **two sub-**

groups,[33] which can at least be distinguished as two single lines in the spectroscopic apparatus. An additional result that follows from Sommerfeld's theory is that the difference of the two frequencies for all hydrogen doublets must have the same constant value, and this has been confirmed experimentally. The theoretical value of the **frequency difference of the hydrogen doublet** is in exact agreement with the value actually observed.

In the case of helium, for which the single lines of the sub-groups are not situated quite so closely together as with hydrogen,[34] even the sub-groups could be resolved in apparatus of high resolution, and the experimental investigations with helium gave a brilliant confirmation of Sommerfeld's theory. All the lines appeared exactly as the theory of fine structure predicts. Moreover, the agreement of the Bohr-Sommerfeld theory with experiment is so complete that it is possible, **solely from spectroscopic measurements,** to calculate the elementary quantum of action and the two fundamental magnitudes of the electron theory, viz. the elementary quantum of electricity and the mass of the negative electrons. In order to evaluate these three fundamental constants of modern physics, we only need to measure the ordinary spectrum lines of hydrogen, determine the supposed deviations of the helium spectrum from the theory as previously described, and finally find the frequency difference of the hydrogen doublet.

At the present time, in theoretical physics, the

[33] The manifold of spectral lines is dual in Bohr's theory, and in Sommerfeld's theory quadruple Hence in Sommerfeld's theory every Bohr line appears as a dual manifold.

[34] There is, of course, also a theoretical reason for this.

best method of proceeding to obtain a **very accurate** evaluation of the **atomistic constants** is to **combine** a direct determination of the **elementary quantum of electricity** with the spectroscopic methods. In the determination of the electrical quantum a small charged particle of matter, preferably in the form of an oil droplet, is observed by a microscope under the double influence of its own weight and of a vertical electrical field acting upwards. When the particle is suspended, its weight must be just as large as the electric force, which, for a given electric field, is in its turn proportional to the charge of the particle. Now the weight of the particle can be evaluated from its velocity of free fall according to a definite formula, and thus the charge itself can also be measured. If the charge is sufficiently small it is found to be the elementary quantum of electricity or an integral multiple of it. In this way **Millikan** was able to determine the electrical quantum very accurately, with a limit of error of one per mille.[35]

When the magnitude of the elementary quantum of electricity is known, we can easily find the **mass of an atom of hydrogen**, for on the basis of measurements initiated by **Faraday** we know very accurately the quantity of electricity transported by a gram of hydrogen in electrolysis.[36] We thus know the ratio between the charge and the mass of an ionized

[35] Millikan obtained in this way the value of $4{\cdot}774 \cdot 10^{-10}$ electrostatic units (cf. notes 7 and 8 in the third lecture). If the particles have a radius less than about one-thousandth of a millimetre, however, the method of Millikan is found to be no longer applicable.

[36] A gram-atom of a uni-valent element is deposited in one second by a current of 96494 amperes. By a gram-atom of an element we understand as many grams as are represented by the atomic weight (e.g. a gram-atom of hydrogen weighs 1·008 grm., etc.).

hydrogen atom. Since it must be assumed that a charge of one elementary quantum of electricity is attached to an ionized hydrogen atom, it follows that if we know the value of the elementary quantum we also have the mass of an atom of hydrogen. In this way we find that to within 1 per cent. a quadrillion atoms of hydrogen weigh 1·66 grm.[37]

But the **mass of the negative electron** is also accurately known, as soon as we have first found the exact value for the electrical quantum. This follows from the fact that the specific charge of negative electrons is known both from the deflection of cathode rays and also (according to Bohr's theory) from the deviations that exist between the helium and the hydrogen spectra.[38] Finally, if we know the accurate values for the elementary quantum of electricity and the mass of the negative electron, the **elementary quantum of action** can also be calculated from the formula for Rydberg's constant, with an accuracy of about one-half per cent.[39]

The methods just given are probably the most accurate, but by no means the only ones. A series of phenomena that permit of a determination of atomistic constants has already many times been mentioned, e.g. the deflection of the alpha-rays, scintillations, the photoelectric effect, Stefan's law,

[37] The mass of a hydrogen atom is $1{\cdot}662 \,.\, 10^{-24}$ grm. Closely related to this constant is the so-called Loschmidt number, which expresses the number of atoms of any element contained in a gram-atom of that element. The Loschmidt number thus represents the number by which we must divide the atomic weight of an element expressed in grams, in order to obtain the actual mass of its atom. The magnitude of the Loschmidt number is 606 thousand trillions.

[38] The mass of a negative electron is $9{\cdot}00 \,.\, 10^{-28}$ grm.

[39] The accurate value of the elementary quantum of action is $6{\cdot}545 \,.\, 10^{-27}$ erg. secs.

and Wien's displacement law, whilst the mass of the atom can also be calculated from Brownian movement and from other molecular phenomena of fluctuation. Nevertheless, we obtain the same values for the fundamental magnitudes of physics by all these methods, within the limits of error.[40]

If it can be regarded as an argument for the existence of our external world that the sensations of sight, hearing, and touch all lead us to infer the existence of the same objects, then theoretical physics has **certain proof of the real existence of electrons, of atoms, and of the elementary quantum of action** in the fact that the characteristic constants of atomistics, as obtained by fundamentally different methods, nevertheless always have **the same values.**

[40] Compare the author's elementary article "Die Loschmidt sche Zahl und die modernen Methoden ihrer Bestimmung" ("Die Naturwissenschaften", *9*, 1921, pp. 180-184).

V. THE THEORY OF THE CHEMICAL ELEMENTS

THE electron theory and the quantum theory have led to a complete transformation not only in physics, but they have produced quite new foundations also for the science of **chemistry**. These theories have brought the solution to the oldest and fundamental problem of theoretical chemistry, the **problem of the chemical elements.**

Even at the beginning of the nineteenth century, by virtue of **Dalton's theory** (discussed in the third lecture), this problem had been recognized as synonymous with the question of the **types of atoms.** At that time chemists were only acquainted with some thirty chemical elements. As a result of the discovery of the many rare earth metals[1] and the rare gases,[2] the number of known elements increased until at the end of the nineteenth century there were about eighty, and since the discovery of the manifold so-called radio-elements the number has increased to over a hundred.

Until comparatively recently the **atomic weight** of an element had been regarded as its most important and distinctive property, and for this reason it was

[1] The discovery of the rare metals was not possible until after the introduction of spectrum analysis.

[2] The rare gases were discovered in the atmosphere by Ramsay towards the end of the nineteenth century.

customary to arrange the elements in a **series of increasing atomic weight,** of which the first member was **hydrogen** and the last **uranium.** Following the usual practice, if we choose the sixteenth part of the mass of an atom of **oxygen** as our **unit of atomic weight,**[3] then the series begins with the number 1·008 and ends with the number 238·2.

In this series, however, it is found that the atomic weight increases by no means uniformly. The differences of atomic weight between two successive elements of the series are far from being the same, and hence we could omit one or other of the elements from the series, without the resulting gap being obvious. It was thus nothing short of an epoch-making advance, when in the year 1913 a physical discovery led to the detection of a **natural series of the chemical elements.** This discovery, which we owe to the English physicist **Moseley,** who died quite young, stands in close relation to the foundation of **Röntgen-spectroscopy** (which is likewise the work of Moseley).

The spectroscopy of Röntgen rays is based on the experimental fact that bodies on which Röntgen rays impinge emit so-called **secondary Röntgen rays,**[4] the frequencies of some of which depend only on the chemical nature of the body struck by the incident rays.[5] This so called **characteristic radiation** is thus composed of rays of definite wave lengths, which are

[3] Compare note 1 in the third lecture.

[4] The emission of secondary Röntgen rays is an analogous phenomenon to that of optical fluorescence; the difference between the two phenomena is only of a quantitative nature. Cf. note 13 in the fourth lecture.

[5] This was first discovered by Barkla in 1905, but he was only able to determine the hardness of the rays, and not their wave lengths.

just as characteristic of the elements contained in the radiating body as are the optical spectra of these elements. The discoveries of Laue (referred to in the first lecture) and Bragg made it possible to measure the wave lengths of the characteristic radiation exactly, and Moseley first succeeded in doing this for a variety of elements by means of an ingenious method, in which the Röntgen spectra were simply photographed. In this way Moseley discovered in the Röntgen spectra of all the solid elements examined two series, which are distinguished by the names **K-series** and **L-series**; the K-series lies in general about three octaves higher than the L-series. Later, a third series was discovered for elements of very high atomic weight. It is known as the **M-series**, and is situated about two octaves lower down than the L-series.[6]

When Moseley photographed the Röntgen spectra of different elements and compared them with each other, he made the surprising discovery that the elements can be arranged in a natural series according to their Röntgen spectra. Within this series the **displacement** of the individual lines in the direction of increasing frequency occurs with remarkable **regularity** from element to element, and with an **exactness** such that every **gap** in this series is immediately revealed by too large a jump. Moreover, the sequence of the elements in this natural series is almost exactly the same as in the series obtained by arranging the elements in the order of increasing atomic weight. But whereas the increase in the atomic weight is by no means a regular one, the displacement of the Röntgen lines within the natural

[6] The M-series was discovered by Siegbahn.

series occurs with wonderful precision. (In Fig. 3, due to Moseley, and of historical importance, we see the two strong K-lines of the successive elements calcium, titanium, vanadium, chromium, manganese, iron, cobalt, nickel, copper, and zinc, which appears in the alloy with copper as brass. Scandium, which occupies the place between calcium and titanium, was at that time not available owing to its rareness and is thus missing in the figure. It should also be noted that in the spectrum of brass the lines of copper are present along with those of zinc, which clearly proves that Röntgen spectra are a property of the atoms themselves. For the same reason we can observe the lines of both cobalt and nickel in the spectrum of cobalt, with which almost always traces of nickel and iron are associated.)

Since the atomic weight increases from place to place (even though not regularly) in the natural series of the elements (with few exceptions), we can fill in the existing gaps on the basis of the known atomic weights, in so far as they refer to elements that are known, but for which Röntgen-spectroscopic photographs are not available.

The natural series of the chemical elements obtained in this manner comprises **ninety-two places** between hydrogen and uranium, and including these. It is represented by Table I, in which, in addition to the so-called **atomic number** (classification number) which determines the number of the place in the series, and in addition to the conventional names of the elements, the chemical symbols and the atomic weights relative to O = 16 are also recorded. As shown in the table, there are four instances in which an element of higher atomic weight precedes one of lower atomic weight (viz. argon precedes

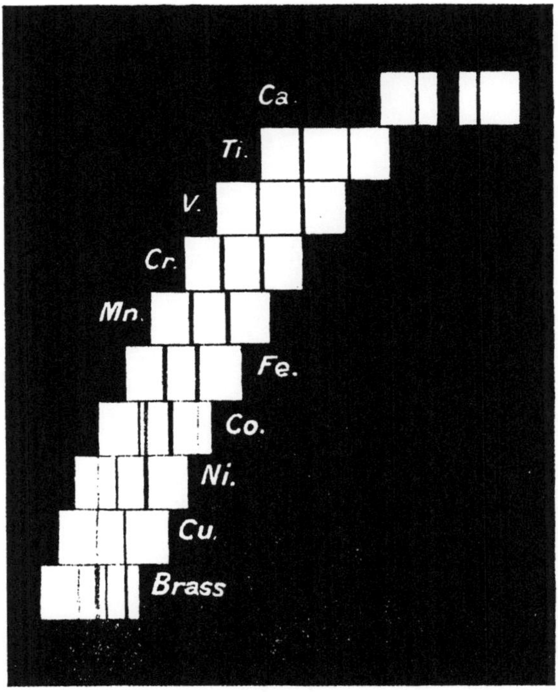

FIG. 3

[*To face page* 100

TABLE I

THE NATURAL SEQUENCE OF THE CHEMICAL ELEMENTS

Atomic Number.	Element.	Symbol.	Atomic Weight.	Atomic Number.	Element.	Symbol.	Atomic Weight.
1	Hydrogen	H	1·008	47	Silver	Ag	107·88
2	Helium	He	4·00	48	Cadmium	Cd	112·40
3	Lithium	Li	6·94	49	Indium	In	114·8
4	Beryllium	Be	9·1	50	Tin	Sn	118·70
5	Boron	B	11·0	51	Antimony	Sb	121·77
6	Carbon	C	12·005	52	Tellurium	Te	127·5
7	Nitrogen	N	14·008	53	Iodine	I	126·92
8	Oxygen	O	16·000	54	Xenon	X	130·2
9	Fluorine	F	19·001	55	Caesium	Cs	132·81
10	Neon	Ne	20·2	56	Barium	Ba	137·37
11	Sodium	Na	23·00	57	Lanthanum	La	139·0
12	Magnesium	Mg	24·32	58	Cerium	Ce	140·25
13	Aluminium	Al	26·963	59	Praseodymium	Pr	140·9
14	Silicon	Si	28·111	60	Neodymium	Nd	144·3
15	Phosphorus	P	31·04	61	—	—	—
16	Sulphur	S	32·06	62	Samarium	Sa	150·43
17	Chlorine	Cl	35·46	63	Europium	Eu	152·0
18	Argon	A	39·88	64	Gadolinium	Gd	157·3
19	Potassium	K	39·10	65	Terbium	Tb	159·2
20	Calcium	Ca	40·07	66	Dysprosium	Dy	162·5
21	Scandium	Sc	45·1	67	Holmium	Ho	163·5
22	Titanium	Ti	48·1	68	Erbium	Er	167·7
23	Vanadium	V	51·0	69	Thulium	Tm	168·5
24	Chromium	Cr	52·0	70	Ytterbium	Yb	173·5
25	Manganese	Mn	54·93	71	Lutecium	Lu	175·0
26	Iron	Fe	55·84	72	Hafnium	Hf	—
27	Cobalt	Co	58·97	73	Tantalum	Ta	181·5
28	Nickel	Ni	58·68	74	Tungsten	W	184·0
29	Copper	Cu	63·57	75	—	—	—
30	Zinc	Zn	65·37	76	Osmium	Os	190·9
31	Gallium	Ga	70·1	77	Iridium	Ir	193·1
32	Germanium	Ge	72·5	78	Platinum	Pt	195·2
33	Arsenic	As	74·96	79	Gold	Au	197·2
34	Selenium	Se	79·2	80	Mercury	Hg	200·6
35	Bromine	Br	79·92	81	Thallium	Tl	204·0
36	Krypton	Kr	82·92	82	Lead	Pb	207·20
37	Rubidium	Rb	85·45	83	Bismuth	Bi	209·0
38	Strontium	Sr	87·63	84	Polonium	Po	(210)
39	Yttrium	Y	89·33	85	—	—	—
40	Zirconium	Zr	90·6	86	Niton	(RaEm)Nt	(222)
41	Niobium	Nb	93·5	87	—	—	—
42	Molybdenum	Mo	96·0	88	Radium	Ra	226·0
43	—	—	—	89	Actinium	Ac	(226)
44	Ruthenium	Ru	101·7	90	Thorium	Th	232·15
45	Rhodium	Rh	102·9	91	Protactinium	Pa	(230)
46	Palladium	Pd	106·7	92	Uranium	U	238·2

potassium, cobalt comes before nickel, as is shown clearly in Fig. 3, then tellurium precedes iodine, and finally thorium occurs before protactinium). **Five** of the ninety-two places in the series belong to elements that have **not yet** been **discovered,** so that we now know how many elements there are to be discovered, unless there should happen to be any with an atomic weight higher than that of uranium.

From the standpoint of the modern theory of the atom the natural series of the chemical elements finds a simple interpretation. If the atoms consist of positive nuclei round which the negative electrons revolve, it seems plausible to assume that as we advance from one place to the next in the series the nuclear charge also increases by one elementary quantum of electricity. This assumption is strongly supported by Moseley's discovery that the square roots of the frequencies of the individual Röntgen lines increase linearly with the ordinal (atomic) numbers of the elements,[7] whereas it follows from Bohr's theory on the other hand that with increasing nuclear charge the frequencies must increase in proportion quadratically. (In the spectrum of ionized helium, for instance, the frequencies of the corresponding lines are throughout four times as large as in the spectrum of hydrogen, the nuclear charge of helium being double that of hydrogen.) The **ordinal numbers** of the chemical elements are thus found to be **identical with the nuclear charge numbers,** which state how many positive elementary quanta are contained in the charge of the positive nucleus.

[7] If we plot the ordinal numbers of the elements as abscissæ, and the square roots of the frequencies of a particular line as ordinates, the resulting graph is very nearly a straight line.

Negative electrons revolve round the atomic nucleus, and theoretical physics is faced with the important task of **deriving from the quantum theory the arrangement of the negative electrons surrounding the nucleus.** Here the foundation for a **schematic representation of atom types** seems to have been laid by an investigation published by **Bohr** in the year 1921.[8] According to this work, Bohr is able to deduce from the quantum theory that with increasing nuclear charge number at first **two** electrons must group themselves round the nucleus, then round these other **eight** electrons, and around these again eight electrons. If the nuclear charge number increases still further, a group of **eighteen** electrons is pushed in **between** the two groups of eight, and then a second group of eighteen electrons, and finally between the two groups of eighteen an additional group of **thirty-two** electrons is inserted. The grouping of the negative electrons about the nucleus is represented in Table II.

TABLE II

2					
2	8				
2	8	8			
2	8	18	8		
2	8	18	18	8	
2	8	18	32	18	8

Now since negative electrons tend to group round the positive nucleus according to the above scheme, it follows from this that in general the number of these electrons can **differ** from the nuclear charge number. Only those elements would constitute exceptions to this, for which the

[8] Cf. "Zeitschrift für Physik," Vol. 9, pp. 1-67.

ordinal numbers are equal to the sums of the electron numbers in the horizontal rows of Table II. For those elements with the ordinal numbers 2, 10, 18, 36, 54, and 86, the nuclear charge number will **always** be equal to the number of negative electrons surrounding the nucleus. For the remainder of the elements this will generally not be the case, and the number of negative electrons surrounding the nucleus may be **different** from the nuclear charge number. This will be so on the **grounds of stability,** because the electrons surrounding the nucleus have a tendency to group themselves more or less according to the scheme already given.

But if the nuclear charge number is not equal to the number of electrons surrounding the nucleus, an **excess** of positive or negative electricity must result, and in such a case the **atom as a whole** must therefore possess an **electric charge.** Thus **electrical forces** either of attraction or of repulsion must proceed from such an atom, the nucleus of which is not completely neutralized by the surrounding negative electrons. These electrostatic forces that proceed from an atom which is not completely neutralized are interpreted on the modern theory of the atom as **chemical affinity** or chemical **valency.**

The elements with the ordinal numbers 2, 10, 18, 36, 54, and 86 must accordingly be incapable of every manifestation of chemical affinity, and in point of fact, the elements with these ordinal numbers are none other than helium, neon, argon, krypton, xenon, and niton (emanation). These are well-known to chemists as the **rare gases,** and their behaviour from a chemical point of view is entirely

passive, in as much as they form **no chemical compounds** whatsoever.

Those chemical elements which have a nuclear charge number higher by one than the rare gases must readily part with a surplus electron, and consequently their atoms assume a total positive charge of one elementary quantum. These elements (the so-called **alkali metals**) are **electropositive** and **monovalent.** Similarly, the elements which immediately precede the rare gases in the natural classification are electronegative and mono-valent; these are the so-called **halogens.** The elements which follow immediately after the alkali metals, the so-called alkaline earths, are again electropositive and di-valent, whilst the elements that immediately precede the halogens in the natural classification are electro-negative and di-valent, and so on.

Thus the modern theory of the atom gives a satisfactory explanation of a remarkable regularity which, as long ago as 1870, was discovered simultaneously, yet independently, by the German chemist Lothar **Meyer** and the Russian chemist **Mendeléeff.** This regularity is the so-called **chemical periodicity.** Meyer and Mendeléeff found that in a series of the elements arranged in the order of increasing atomic weight, certain properties recur periodically, and with such regularity that it was even possible for Mendeléeff to predict the properties of three elements not actually discovered till later (scandium, gallium, and germanium), from the known properties of boron, aluminium, and silicon.

The chemical periodicity manifests itself in the fact that within the natural series or classification of the elements **eight groups** are distinguishable. With each of these eight groups is associated a different **chemical valency.** In general, an atom of an element combines with as many atoms of a halogen as correspond to its group number (e.g. NaCl, $MgCl_2$, $AlCl_3$, $SiCl_4$, PCl_5, SF_6). Conversely, an atom of the fourth group combines with four hydrogen atoms, one of the fifth group with three, one of the sixth group with two, and one of the seventh group with one hydrogen atom. This division of the natural series of the elements into eight groups appears, then, according to Bohr's latest theory, to be determined by the fact that (from Table II) the outermost group of electrons must always comprise eight when fully occupied.

The **periodic system of the elements** is represented in Table III, with the addition of the ordinal numbers to the chemical symbols. As the Table shows, the natural series of elements is broken up at the rare gases into a "fore-period" of two elements, into two "small periods" of eight elements each, into two "large periods" each with eighteen elements, and into a "very large period" of thirty-two elements. This is succeeded by a new period which, however, appears to break off at uranium in the sixth group.

Since in a large and in the very large period each value of the valency occurs twice, it is possible to distinguish between two sub-groups within each of these groups; in the same sub-group we have to fit in those elements which are situated equally far from the rare gases in the natural series of the elements. Within the first group, for instance, copper, silver, and gold constitute a sub-group, and within the second group, zinc, cadmium, and mercury. The three "**triads**" (iron, cobalt, nickel; also ruthenium, rhodium, palladium; and finally the three "platinum metals" osmium, iridium, platinum) are

TABLE III

	I. a.	I. b.	II. a.	II. b.	III. a.	III. b.	IV. a.	IV. b.	V. a.	V. b.	VI. a.	VI. b.	VII. a.	VII. b.	VIII.	O.
1	1H															2He
2	3Li		4Be		5B		6C			7N		8O		9F		10Ne
3	11Na		12Mg		13Al		14Si			15P		16S		17Cl		18A
4	19K	29Cu	20Ca	30Zn	21Sc	31Ga	22Ti	32Ge	23V	33As	24Cr	34Se	25Mn	35Br	26Fe 27Co 28Ni	36Kr
5	37Rb	47Ag	38Sr	48Cd	39Y	49In	40Zr	50Sn	41Nb	51Sb	42Mo	52Te	43—	53I	44Ru 45Rh 46Pd	54X
6	55Cs	79Au	56Ba	80Hg	57-71 rare earths	81Tl	72Hf	82Pb	73Ta	83Bi	74W	84Po	75—	85—	76Os 77Ir 78 Pt	86Em(Nt)
7	87—		88Ra		89Ac		90Th		91Pa		92U					

remarkably placed in the periodic system. The succession of the fifteen elements of the rare earths is also singular, but their unique position is explained by Bohr's most recent theory.

Elements belonging to the same group or subgroup show quite similar properties, and the periodicity is evidenced not only in the valency but also in many physical properties.[9] The nature of elements not yet discovered can also be predicted from the periodic system. The elements of ordinal numbers 43 and 75 must resemble manganese, 61 must be a rare earth element, 85 a halogen, and 87 an alkali metal.

Attention has already been drawn to the fact that, on quantum-theoretical grounds and for reasons of stability, the atoms of all the elements with the exception of the rare gases show the tendency to assume a surplus of positive or negative electricity. In consequence of the electrostatic attraction proceeding from the charged atoms, they strive to neutralize this surplus charge by **combining** with other non-neutral but **oppositely charged** atoms **to form molecules.**

Such a combination between oppositely charged atoms is called a **heteropolar combination.** But since every atom is composed of individual positive and negative electrons, it can also happen that two neutral atoms exert an attraction on each other, provided that oppositely charged electrons (i.e. protons and electrons) of the two atoms are brought sufficiently

[9] This is true particularly for the atomic volume (quotient of atomic weight and density), and also for the melting-point, the coefficient of expansion, the compressibility and many electrical, magnetic, optical, and thermal properties.

near to each other. In such cases we have to do with so-called **homœopolar** combinations.

The most important examples of homœopolar combinations are those between **atoms of the same element.** It follows from the measurement of gas density and from Avogadro's law [10] that in the gaseous state the molecules of **hydrogen, nitrogen, oxygen** and the **halogens** are entirely **diatomic,** as are also the molecules of selenium and tellurium. Molecules consisting of **eight atoms** have been detected in the case of **sulphur,** and molecules composed of **four atoms** in the case of **phosphorus and arsenic** at not too high temperatures.[11] For carbon it must be assumed that in the solid state its molecules contain at least twelve atoms.[12] **Ozone,** a modification of oxygen, is **triatomic.** Determinations of the vapour densities of the metals of the first group of the periodic system have without exception led to the result that with them a formation of molecules does not occur, or, as it would be expressed in the

[10] The diatomic nature of hydrogen, oxygen, nitrogen, and of the halogens results from an investigation of the volume relations for gaseous reactions, on the basis of Avogadro's law (cf. note 15 in the second lecture). For example, a litre of hydrogen and a litre of chlorine combine to form two litres of hydrochloric acid gas. Thus according to Avogadro's law just as many hydrochloric acid molecules must be present after the reaction, as is represented by the sum of the previously existing hydrogen and chlorine molecules. But since the molecules of hydrochloric acid must consist of (at least) two atoms, a chlorine atom and a hydrogen atom, it follows that the molecules of these two elements must also each be composed of two atoms.

[11] Both phosphorus and arsenic vapours are diatomic at high temperatures.

[12] This was concluded from the fact that in the oxidation of amorphous carbon by potassium permanganate mainly mellitic acid results, in the molecules of which are contained twelve atoms of carbon.

language of chemistry, their molecules are uni- or **mon-atomic.**[13] It seems self-evident that all the rare gases are likewise mon-atomic, in view of their complete chemical inertness.

The **structure of crystals** must also be regarded as a particular case of the formation of molecules, and one about which physics has been able during the last decade to obtain definite information. From the **Laue photograms** mentioned in the first lecture, two English physicists, **W. H. and W. L. Bragg**, were

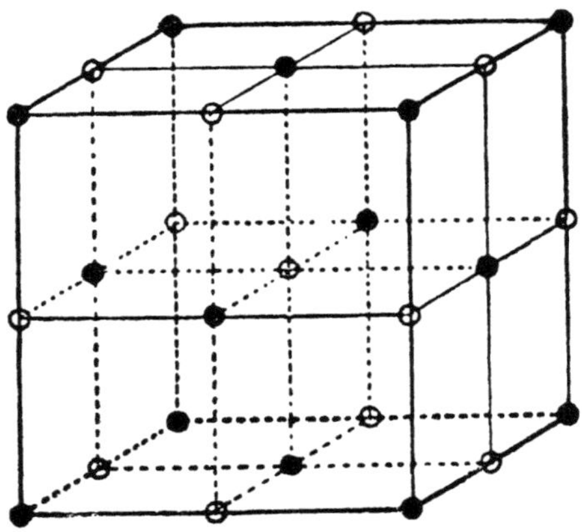

FIG. 4.

able in 1913 to draw convincing conclusions as to the constitution of different crystals, as for instance in the case of **rock salt**, which is a combination of equal numbers of atoms of sodium and chlorine. They found that the structure of rock salt crystals is exceedingly simple; the sodium and the chlorine atoms are arranged alternately in a **cubic lattice** and at equal distances. In Fig. 4, which must be

[13] Such determinations of vapour density have been successful, particularly in the case of mercury and cadmium.

supposed to extend in all directions, the sodium and the chlorine atoms are distinguished by small light and dark circles. The structure of other crystals is similar. An interesting example of a crystal built up of only a single element is furnished by **diamond** (Fig. 5), in which any four neighbouring carbon atoms can always be combined to form a tetrahedron.

Crystals represent single giant molecules, in which, however, owing to the perfectly regular arrangement, the different kinds of atoms are at the

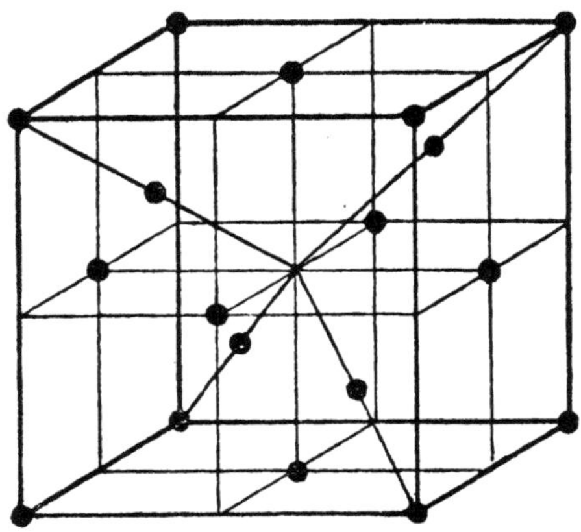

FIG. 5.

same time combined with each other in very simple numerical relations, so that a simple formula and a corresponding definite molecular weight can be assigned to the crystal compound.

With regard to the **nuclei of the atoms,** they must obviously also be composed of positive and negative electrons. Since the mass of the negative electrons is quite inappreciable as compared with the mass of the positive electrons, it follows that the atomic weight of an element relative to hydrogen represents nothing else than the number of positive electrons contained in the nucleus. But this number is found to be about twice as large or still more than the nuclear charge number of the elements. It must be concluded from this that there are **also negative electrons in the nucleus**, and about half as many as there are positive ones. The charges of the positive electrons contained in the nucleus are only partially compensated by the negative electrons present in it, so that the nucleus of the atom is left with a net positive charge.

On theoretical grounds it is very probable that the structure of the atomic nuclei is also governed by quantum relations, but for the present we have no certain experimental evidence of this.[14] The most important phenomenon known to take place **within the nucleus of the atom** is the **spontaneous disintegration of the nucleus**, and the phenomena dependent upon this are grouped together under the name **radioactivity.** The disintegration consists of a **splitting off** of components of the nucleus, and since the **nuclear charge is altered** by such disintegration, it follows that elements which emit α- or β-rays must undergo consequent **transformation.**

As a result of α-radiation the chemical ordinal number will be reduced by two, because the α-particles have a charge of two positive elementary quanta. On the other hand, the β-particles each have a charge of one negative elementary quantum, so that the ordinal number will be increased by one in consequence of β-radiation. In addition, α-particles have a mass equal to that of four hydrogen atoms, whereas the mass of a β-particle is small compared with the mass of an atom of hydrogen.

[14] Compare the last two paragraphs of this lecture.

α-**radiation** thus **displaces** an element in the **periodic system through two places to the left,** and at the same time **decreases the atomic weight by four units.** β-**radiation** displaces an element through **one place to the right** in the periodic system, without appreciable alteration of the atomic weight. In this connection we see from Table III that a displacement through six places to the left is equivalent to a displacement through two places to the right. The displacement law of the transformation of the chemical elements was established in 1913 by **Soddy** and **Fajans** simultaneously, but independently of each other. The displacement law provided for the radioactive transformations a systematic basis, the lack of which had been greatly felt up to that time. It enabled scientists to determine the chemical nature of all the products of transformation, to fill in previously existing gaps, and finally, it also led to the discovery of new radioelements.[15]

Two coherent **transformation series** of elements are at present known in physics. One of these series starts off from **uranium** and branches off later into the **radium series** and the **actinium series.** The parent element of the other series is **thorium.**

The **uranium series** is represented in Table IV, in which are included, alongside the chemical symbols of the radioelements,[16] their respective group numbers (in terms of the vertical rows of Table III), atomic weights, and half-value periods,[17] all in

[15] Of particular interest in this respect was the discovery of protactinium by Otto Hahn and Lise Meitner, and by Soddy.

[16] Apart from the symbols shown in Table I, the abbreviations signify: Io = Ionium, Rd = Radio.

[17] A few of the transformations shown in Table IV have not yet been confirmed experimentally, but must be presumed on theoretical

TABLE IV

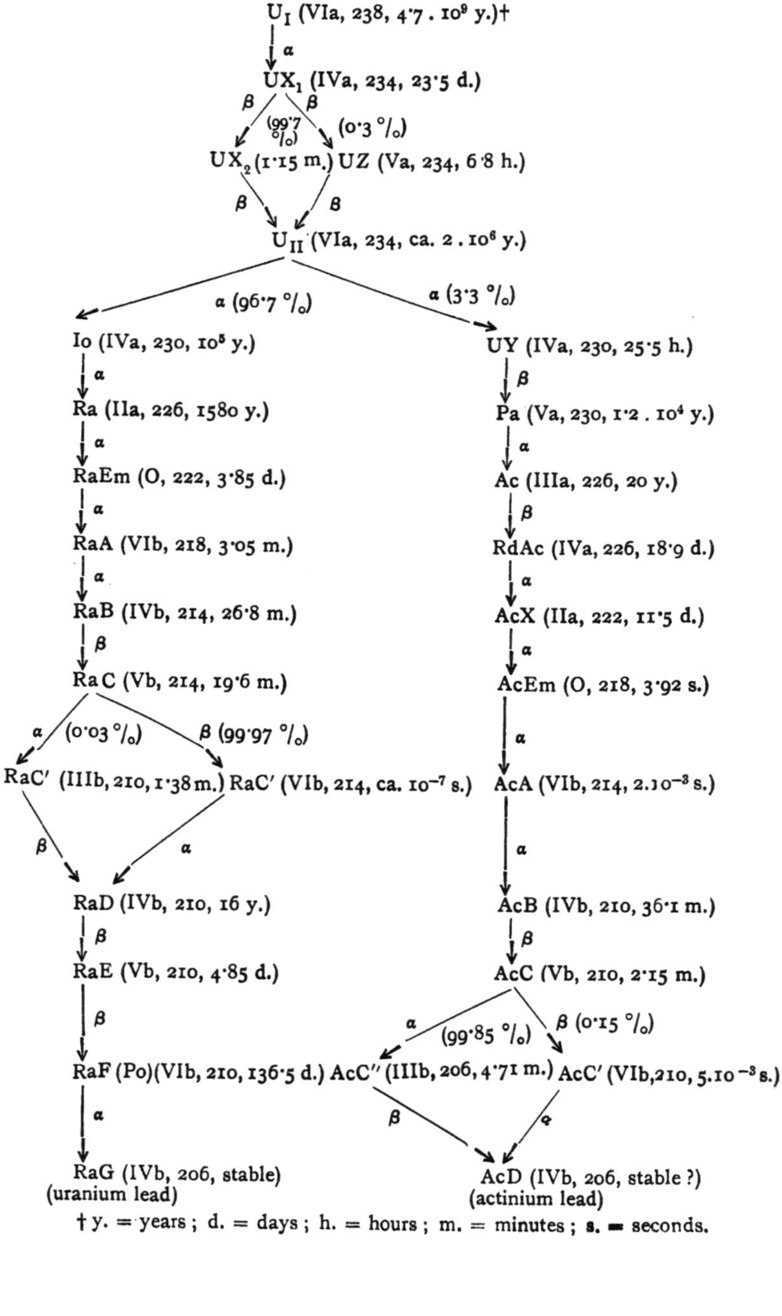

† y. = years ; d. = days ; h. = hours ; m. = minutes ; s. = seconds.

brackets. The empirical confirmation of the group displacement laws can be readily seen from the table. For example, radium is produced from Uranium I after three α-transformations and two β-transformations, and is, in point of fact, displaced through ($3 \times 2 - 2$) or four groups to the left as compared with uranium, so that it occupies the same group as barium, and at the same time its atomic weight is twelve units less than that of uranium. Again, the emanation occupies a place two groups to the left of radium, from which it is produced by an α-transformation, or, what amounts to the same thing, it is displaced six places to the right as compared with radium. As a matter of fact, the emanation has the properties of a rare gas, whereas radium is akin to the alkaline earths. After five α-transformations and four β-transformations, lead is produced from radium, and we actually find that it is situated ($5 \times 2 - 4$) or six places to the left of radium, i.e. from group IIa to group IVb. The diminution of atomic weight amounts to five times four, or twenty units.

Up to the present it has not been possible to trace the transformation series beyond lead, and although we know that the stability of lead is of a very high order, it is conceivable that the series may not actually end at this element. Hitherto only two cases have been found in which radioactivity has been shown by elements of a lower ordinal number than that of lead; the elements concerned are the alkali

grounds. Of particular interest is the branching of the transformation series. Only a small fraction of the atoms (about 0·03 per cent.) of RaC are transformed into RaC″, whereas the great majority of them undergo transformation into RaC′. The series also branches at AcC.

metals **potassium** and **rubidium**, both of which emit β-rays.

In radioactive processes the disintegration of the nucleus takes place spontaneously, and cannot be influenced by any outside agency known to us. It was therefore a discovery of the first rank, when **Rutherford** in 1919 first succeeded in **breaking up an element by artificial means**, the more so, seeing that the element was one of low ordinal number.

Rutherford's discovery was based on observations of the **passage of α-particles through gases.** When an α-particle collides with an atomic nucleus of the gas traversed, the α-particle is not alone in receiving a deflection. According to Newton's well-known Third Law of Motion (the principle of the equality of action and reaction), the **nucleus of the atom with which the α-particle collides must also** be subject to **recoil**, and it will of course be projected farther, the lighter it is. Thus when **Marsden** passed α-particles through **hydrogen**, he was able to detect **scintillations** (see the third lecture) on a zinc sulphide screen even at a distance of more than 80 cm. from the source of the α-rays, although the α-particles themselves only penetrated a distance of about 24 cm. Later, Rutherford showed experimentally that the particles of which these rays of such long range consist have the same specific charge as positive electrons, i.e. as hydrogen nuclei. From the chemical symbol for hydrogen these rays of long range were therefore called **H-rays.**

Soon after the discovery of the H-rays, Marsden and his collaborator **Lantsberry** made the surprising observation that a nickel foil coated with radium-C, and surrounded by ordinary air, produced scintillations at a distance much greater than the range of the α-particles emitted by radium-C. When **Rutherford** followed up this at first puzzling phenomenon, he made the important discovery that the number of scintillations depended on the substance that occupied the space between the nickel foil and the scintillation screen. The scintillations did not occur when the intervening space was empty or filled with carbon dioxide or oxygen instead of with air. However, when Rutherford replaced the air by pure nitrogen, the number of scintillations increased.

This fact led him to the conclusion that the particles of long range could only have arisen from a **disruption of the nitrogen nucleus.** In the bombardment with α-particles this disruption is achieved by artificial means, whereas in radioactive phenomena the disintegration of the atomic nuclei occurs spontaneously. A closer investigation of the long range rays originating from the disruption of nitrogen nuclei showed definitely that these rays actually consist of **hydrogen nuclei.**[18] In later experiments Rutherford was able, by means of α-rays, to disrupt also the atoms of boron, fluorine, sodium, aluminium, and phosphorus, with accompanying liberation of H-rays.[19]

[18] On closer investigation, Rutherford thought that it was also necessary to conclude that nuclei of an unknown element are also contained in the rays, characterized by an atomic weight 3 and an ordinal number 2. Rutherford called this element X_3. However, later work indicates that this assumption was not in harmony with fact.

[19] Rutherford found that H-rays are not liberated from the nuclei of the elements C, O, S, Ca, Ti, Mn, Cu, Sn. This is probably connected with the fact that most of these elements have atomic weights divisible by 4, so that their nuclei are built up from helium nuclei only. The result of Rutherford's experiments is at present uncertain in the case of the elements Li, Be, Mg, Si, Cl, and K.

Rutherford's sensational discoveries have the semblance of being a direct experimental confirmation of a bold hypothesis that was formulated by **Prout** as far back as the year 1815, and according to which the atoms of all elements are supposed to be built up **from atoms of hydrogen,** the lightest of all the elements. Prout based his assumption chiefly on the suggestion that all **atomic weights,** relative to hydrogen, are **whole numbers.** However, this was later found to be in disagreement with fact. When atomic weight determinations were improved upon, **deviations from the integral values** were detected for many of the elements; for example, this was particularly the case with **chlorine,** the atomic weight of which was found to be 35½, apparently in complete disagreement with the hypothesis of Prout. Thus, for the time being, Prout's hypothesis was thrust into the lumber room of science. This disagreement, which has been particularly embarrassing since the introduction of the electron theory, has nevertheless recently found a complete solution as a result of the significant discovery of **mixed elements.**

The modern theory of the atom gives a simple explanation of the existence of mixed elements. According to this theory, the **nuclear charge** of an atom is determined by the **difference** between the numbers of positive and negative electrons contained in the nucleus, whereas the **atomic weight** depends only on the **total number of the positive electrons** (since the mass of the negative electrons is insignificant in comparison with the mass of the positive ones). Hence two atoms with the same nuclear charge can readily have different masses, and the elements to which these atoms belong must thus have **different atomic weights,** in spite of the fact that they occupy the same position in the periodic system and therefore exhibit the **same chemical properties.** Elements of this kind are called **isotopic** elements, since in Greek "*isos topos*" means the same place, and these elements actually occupy the same place in the periodic system. Furthermore, the necessity of the idea of isotopy was rendered evident by the fact that only nine places were available in the periodic system for some forty radioactive elements.

Isotopes must obviously show the same chemical and practically the same physical behaviour.[20] Only those properties can be different which are determined by the mass of the nucleus. Isotopic elements **cannot** therefore **be separated from each other by chemical methods,** and in this we have an explanation of the occurrence of **mixed elements,** which are none other than a mixture of isotopes. Owing to the perfect agreement in the behaviour of its constituents, such a mixture has the appearance of being a pure element.

A **separation** or at least an **analysis** of mixed elements appears to be possible, however, by methods which are based on the difference in the masses of the nuclei, and the method of **positive ray analysis** devised by **J. J. Thomson** is founded on this idea. These positive rays occur in Geissler tubes at very low pressure; they are **deflected** both in a **magnetic** and in an **electric field,** and by a suitable combination of two such fields they describe **parabolas,** the curvature of which depends on the mass of the positive ray particles. But if the gas atoms have not all the same mass owing to the gas

[20] The very slight difference is particularly noticeable in the case of the so-called rotation spectrum.

being a mixed element, then the mixed isotopes must give rise to **separate parabolas.** By measuring these parabolas it is thus possible to evaluate the atomic weights of the isotopes in the mixture.

Examples of isotopy were first detected with **lead.** It was found that lead has a different atomic weight according as to whether we are dealing with **ordinary** lead or with lead that has been obtained from **uranium** minerals or from **thorium** minerals.

By means of positive ray analysis **Thomson** and **Aston** first showed that the rare gas **neon,** for which the atomic weight 20·2 had been found, was in reality a mixed element containing two isotopes with the **integral atomic weights** 20 and 22. In continuation of this work, **Aston** in 1919 established the much more important fact that **chlorine** is also a **mixed element,** containing **two isotopes** with the integral atomic weights 35 and 37. The apparently uniform admixture of the two chlorine gases (in the ratio of about three to one) in nature finds a ready explanation on the assumption that the earth was originally in the gaseous state.

Aston and other scientists have also been able to prove that numerous other elements are in reality mixed elements; thus **krypton** was found to contain at least six isotopes, **xenon** as many as nine, and **mercury, lithium, boron, magnesium, silicon, argon, potassium, iron, nickel, selenium, bromine, rubidium,** and **antimony** were all found to be mixed elements. On the other hand, no evidence of the existence of isotopy was found with the elements hydrogen, helium, beryllium, carbon, nitrogen, oxygen, fluorine, sodium, aluminium, phosphorus, sulphur, arsenic, iodine, and cæsium. In consequence of the occurrence of the numerous radioelements, the phenomenon of isotopy is exemplified in all the known positions in the periodic system of the elements from thallium upwards.

Radioactivity and the phenomena of isotopy represent the main empirical evidence at present available for the successful investigation of the **inner structure of the nuclei of the atoms.** The **physics of the nucleus** is certainly the branch of physics that constitutes the legacy of the immediate future, and some important work by **Lenz** would appear to be of fundamental significance for a future theory of the atomic nucleus.

From the principle which, according to Bohr, determines the permissible orbits in the hydrogen atom, it follows quite generally that the negative electrons revolving round a positive nucleus describe orbits, the linear dimensions of which approximate to the order of magnitude of a thousand millionth of a centimetre. If, on the other hand, we imagine positive electrons revolving round a negative centre, then according to Bohr's principle, as Lenz has shown, the linear dimensions of the orbits would have to be about a **thousand times smaller** than in the previous case, since the mass of positive electrons is roughly a thousand times larger than that of negative ones. We can imagine aggregates to be formed on a quantum-theoretical basis, and in these we can conceive of positive electrons revolving in orbits and bound by negative electrons. These aggregates would necessarily have linear dimensions about a thousand times smaller than those of the atoms themselves. Aggregates of this kind, then, in which the positive electrons are in the majority, could evidently represent the **nuclei of elements.** In virtue of the electrostatic attraction associated with them, they would endeavour to become neutralized, and this would be accomplished by negative electrons attaching themselves to the nuclei. From what has already been said, it is clear that these negative electrons would revolve round the central aggregates at distances about a thousand times larger than the linear dimensions of the nuclei themselves. By surrounding itself with revolving electrons, the **nucleus** thus **develops into an atom.** On quantum-theoretical grounds, however, these negative electrons which revolve round the nucleus will only be able to arrange themselves in a **stable** manner for quite definite group numbers, and hence the number of revolving negative electrons will not in general be equal to the nuclear charge number. Generally speaking, moreover, atoms that have developed in this way will not yet be neutral; they will themselves still possess a **residual charge,** which will only become **completely neutralized** by the **formation of molecules.**

Thus the **nucleus,** the **atom,** and the **molecule** represent the three **main stages** in the **building up of matter,** under the influence of **electrostatic forces** and by **quantum relations.**

VI. THE THEORY OF RELATIVITY

JUST as the conception of matter lost its original meaning as a result of the electron theory, so another fundamental conception in philosophy, that of time, has been completely changed by the **theory of relativity.** This theory, which was founded in 1905 by **Einstein,** is one of the boldest and at the same time most magnificent creations of the human mind. The further development of the theory of relativity carried out by Einstein led to the solution of the perplexing problem of gravitation, and simultaneously laid bare a wonderful connection between physics and geometry.

The theory of relativity had its origin in a dilemma. From the formulæ of the so-called "classical" physics it follows that, from an **optical** point of view, there must exist something of the nature of **absolute motion,** whereas, from the **mechanical** point of view, the conception of absolute motion had already been recognized as quite **meaningless.**[1] Let us assume

[1] According to Newton's second law of motion, it is only the change in velocity and not the velocity that is determined by the force, and hence the equations of motion have the same form for two systems moving uniformly with respect to each other. By the observation of processes of motion in a system it is thus impossible to recognize whether the system is advancing uniformly or not, and hence it is equally impossible to decide whether the system is or is not in motion. This fact epitomizes the content of the so-called mechanical principle of relativity long known to physicists.

that for one observer light is propagated in all directions at the same rate, and let us imagine a second observer to be **moving** in an arbitrary manner with respect to the first. According to the formulæ of classical physics, the propagation of the light with respect to this second observer cannot possibly take place at the same speed in all directions.

This appeared to lead to the conclusion that the **influence of the earth's motion** on the propagation of light should be of appreciable magnitude, and that it should be possible to detect it by a definite and suitably devised arrangement of an **optical experiment.** But when a difficult experiment of this kind was actually performed by the American physicist **Michelson** in the year 1884,[2] not the slightest trace of such an effect was noticeable. This result was in the highest degree perplexing to physicists, although it must have appeared to them improbable from the outset that absolute motion could occur in one branch of physics, namely, optics, and not in another branch, mechanics. Theoretical physics was thus faced with the difficult task of explaining the **incompatibility** that had thus arisen between classical physics and experience.[3] Its solution required the genius of an Einstein, who exposed a prejudice that had found its way into physics long ago, and had become so firmly rooted that the fact of its existence

[2] For the details of Michelson's experiment reference can be made, say, to the account in Einstein's popular work, "Relativity: The Special and General Theory." Methuen, London, 1920.

[3] Even before the formulation of the theory of relativity (and in some sense preparing its philosophical foundations), Mach and Poincaré showed in their well-known critical and theoretical writings that in many respects classical physics suffered from internal contradictions.

was no longer appreciated. This prejudice was the conception of **absolute time.**[4]

It was perhaps the most radical advance ever made in the annals of theoretical physics when, in 1905, Einstein recognized the invalidity of this quasi-dogmatic conception, and replaced it by a new and truly revolutionary one. He considered that all **statements of time** used in the description of a physical process are only of **relative** significance. According to Einstein, all statements of time depend on the standpoint of the observer describing them, and are thus different for two observers who are in motion with respect to each other. On the other hand, however, it is shown by the experiment of Michelson that there must be a definite **relation** between the different times of the same event noted by two observers in relative motion. Although moving relatively to each other, both observers must nevertheless be **equally justified** in maintaining that, from their point of view, light travels with the same velocity in all directions.

The requirement of the relativity of time, and the postulate of the connection just mentioned between two different relative times, constitute the essence of Einstein's so-called **principle of relativity.** We can also express it by saying that the statements of time by means of which a physical process is described are relative, and dependent on the view-point of the observer; furthermore, that statements of time must satisfy the demand that the **velocity of light,** expressed in terms of these statements, must **always** be **the same** in **all directions.**

It follows as a direct consequence of the principle

[4] It is in large measure due to the powerful influence of Newton that the conception of absolute time was raised to a dogma.

of relativity, that just as there can be no absolute measure of time, so also there can be no absolute measure of length. If a rod has a definite **length** for an observer with respect to whom it is at rest, then that rod will appear **shortened** to an observer who is in motion relatively to the first observer in the direction of the rod's length. A terrestrial observer might find that a rod which is at rest with respect to the earth has a length of one metre; to an observer on the sun, to whom the rod appears to be in motion, it would, if he could compare it with one of his metre rods, seem to be shortened by a 200,000th part of a millimetre. Although this shortening is very small, it is nevertheless sufficient to explain the absence of a positive result in the experiment of Michelson.[5] On the other hand, according to the principle of relativity, a process taking place on the earth appears to proceed **more slowly** as observed from the sun, than it would do to an observer on the earth. Even the idea of **simultaneity** loses its meaning. Two events that appear to an observer to occur simultaneously, no longer seem simultaneous to an observer who is in motion with respect to the previous observer. Also the idea of **shape** becomes a relative one. A body which appears to be spherical to an observer who is at rest with respect to that body will appear as an **oblate ellipsoid of revolution** to a moving observer. A further important conclusion to which the principle of relativity leads is that by combining together velocities we can never obtain a velocity which exceeds that of

[5] Even before the introduction of Einstein's principle of relativity, Lorentz had already made the assumption that a rod in motion appears shortened, in order to explain the experiment of Michelson. For this reason we also speak of the Lorentz contraction.

light.[6] On the principle of relativity, velocities greater than that of light appear impossible. In the theory of relativity the **velocity of light** plays a similar part to **infinity** in mathematics.

Of course all these conclusions appear at first sight to be in the highest degree paradoxical; it is not easy to grasp their meaning nor to become accustomed to them, and in this we have the explanation of the fact that the theory of relativity had to contend with much opposition for some time after its publication. But it soon became evident that the principle of relativity was a discovery of a very far-reaching nature, and even in the same year in which Einstein formulated his principle, he and other scientists had already recognized to what revolutionary consequences the new principle necessarily leads.

As is well known, **Newton's second law of motion** occupies an important place in the foundations of classical physics. This law teaches us that the acceleration of a body always has the same direction as the force producing it, and that the ratio of the force to the resulting acceleration has a value which is independent of the magnitude and of the direction of the force, this ratio being none other than what is called the inertial mass of the accelerated body. It follows from the principle of relativity that this law, although fundamental in classical mechanics, only has an **approximate** and by no means complete validity. It holds only for those motions in which the **velocity**, as observed by the investigator describing them, is **small** in comparison with the velocity of light. But if the motions are so rapid that the

[6] Also the classical formulæ for the composition of velocities, such as the law of the parallelogram of velocities, no longer have any general validity.

velocity relative to the observer is no longer small as compared with the velocity of light, then he must detect **deviations** from the fundamental laws of classical mechanics. In general, the **acceleration** then deviates in its direction from that of the force. But above all, the mass of a body also depends on its **velocity.**[7] When the velocity approaches that of light, the **mass** of the body appears to **increase** very rapidly, and in such a manner that an observer would have to ascribe an infinitely large mass to a body which, with respect to him, was moving with the velocity of light.

Although these remarkable conclusions of the theory of relativity must at first appear paradoxical to any one who is used to thinking in terms of classical physics, they have nevertheless been most beautifully **confirmed by experiment.** In experimental physics there is a well-known phenomenon in which particles with inertial mass actually do move with velocities up to 99 per cent. of the velocity of light. We refer to the phenomenon of the **Beta-rays** discussed in the third lecture, the velocities of which could be determined on the one hand, and on the other, the specific charge of the particles of which they consist, i.e. the ratio between their charge and their mass. Since the charge is independent of the velocity, it follows from the theory of relativity that, in consequence of the increase in mass, the specific charge of the particles must become smaller as the velocity increases. Experiment has shown that this

[7] If a body has a mass m_0 with respect to an observer for whom it is at rest (m_0 is the so-called stationary mass), and if the same body has the relative velocity v with respect to a second observer, then the mass of the body for the second observer is equal to $m_0/\sqrt{1 - v^2/c^2}$, where c is the velocity of light. Thus the mass of the moving body always appears to be increased.

decrease in the specific charge actually does take place, and not only that, but the **quantitative agreement** with the formulæ of the theory of relativity is perfect. The fact of the so-called **fine structure of spectral lines**, discussed in the fourth lecture, constitutes an equally excellent confirmation of the mass formula of the theory of relativity.

In the same year as Einstein published his principle of relativity he also discovered a necessary consequence of this principle in the form of a law of immense significance, the law of the **inertia of energy**. Einstein found that any body which undergoes an alteration of its energy content must at the same time experience an alteration in its mass. If, for instance, the energy content of a body is diminished by thermal radiation, its mass will also be diminished by an amount which is equal to the magnitude of the energy emitted, divided by the square of the velocity of light. If the energy content increases by a definite amount, say by heating or by the absorption of radiation, the mass will also increase by this amount divided by the square of the velocity of light.

It was necessary to conclude from this that inertial mass is inherent in all energy as such, and furthermore, that all mass can only have its origin in energy. **Mass and energy** are thus **identical concepts,** and differ only in a proportionality factor. This factor is equal to the square of the velocity of light, and arises from the difference between the measures used. Mass is associated necessarily with all energy, and energy with all mass.[8] Thus the

[8] Thus, by virtue of its mass, every body is associated with a so-called specific energy, which is equal to the product of its stationary

principles of the conservation of mass and the conservation of energy, the laws of Lavoisier and of Mayer, appear united to a **single principle** by the theory of relativity. In spite of this, both laws appear to play an independent rôle, to a very high degree of approximation. The cause of this lies in the infinitesimal smallness of the alterations of mass connected with observable changes in energy. Even in the case of radium with its relatively enormous development of heat, the annual diminution in mass of one gram of that element, resulting from the emission of heat, would only be about one-half of the millionth part of a milligram, i.e. an amount that would be scarcely detectable.[9]

It is little short of amazing, what an abundance of new laws resulted within a year from Einstein's principle of relativity—laws which completely revolutionized physics. The deeper meaning of the principle of relativity—however, was grasped in the year 1908 by the Göttingen mathematician **Minkowski.** By virtue of the principle of relativity we express a relation between the relative quantities time and the three space co-ordinates used in physics for the statement of a position in space. It was discovered by Minkowski that this relation is

mass and the square of the velocity of light. Compared with the specific energy, the so-called kinetic energy (which has, of course, only a relative significance) is in general vanishingly small; even the kinetic energy corresponding to that of a rifle bullet in full flight would only represent approximately the thousand millionth part of the specific energy of the bullet.

[9] The loss of mass determined by the development of heat is not to be confused, of course, with the loss of radium itself in consequence of the formation of radium emanation. The diminution in mass would be determined for a preparation that had been enclosed in a completely impermeable case.

capable of a **geometrical interpretation.** If we attach a negative sign to the square of the product of the velocity of light and the time, then this negative square and the squares of the three spatial co-ordinates are related to each other in exactly the same way as the squares of four co-ordinates in a **four-dimensional geometry.**[10]

The four-dimensional manifold, which thus appears as a **linking together of space and time,** is known as the **Minkowski world.** With every point in it, with every "**world point**" there are associated **four co-ordinates, three spatial** and **one temporal** co-ordinate. Let us suppose that at a definite time a moving body is situated in a particular place representable by three spatial co-ordinates; then this fact can be expressed symbolically by the fixation of a world point, the spatial co-ordinates of which correspond to the co-ordinates of that place, and the time co-ordinate of which corresponds to the time at which the moving body is situated at that place.

A succession of world points constitutes a "**world line,**" and the motion of every body (considered as being infinitesimally small, or without extension) can thus be symbolically represented by a world

[10] Let us indicate the three space co-ordinates by the symbols x, y, z, the time by t, and the velocity of light by c; on the other hand, let us denote the four co-ordinates in a four-dimensional geometry by x_1, x_2, x_3, and x_4. These four co-ordinates are transformed according to certain formulæ when the four-dimensional system is rotated. In accordance with Minkowski's interpretation of the principle of relativity, these formulæ of transformation retain their validity, when the four magnitudes x_1^2, x_2^2, x_3^2, x_4^2, that occur in them are replaced by the four magnitudes x^2, y^2, z^2, and $-c^2t^2$. The statement that absolute motion cannot exist, appears from Minkowski's standpoint to be equivalent to the statement that no direction can be in any way privileged in the Minkowski world.

line. The standpoint of the observer describing the motion is made manifest solely by the way in which the four-dimensional co-ordinate system is laid in the Minkowski world, one of the axes of that system being the time axis. The **inclination of the world line with respect to this time axis** represents the **velocity** possessed by the moving body from the view-point of the observer. If the world line is **straight,** i.e. if it has the same inclination to the selected time axis throughout, then a **uniform motion** is represented by the world line, that is, a motion for which both the direction and velocity remain permanently the same. Non-uniform motion, on the other hand, is represented by a curved world line, and the **acceleration** is determined by the **curvature** of the world line.

As a result of the principle of relativity established by Einstein in the year 1905, the concept of absolute rest has lost all meaning in physics. But in spite of this, the conception of absolute motion was as yet by no means completely disposed of by the principle, for the equivalence of two systems moving with respect to each other had been thus far confined to the **special case of a motion progressing uniformly.** Moreover, even if no single privileged system of reference can exist according to the principle of relativity, yet there still remains a manifold of privileged co-ordinate systems, which are all in uniform motion with respect to each other. The motion of a body appears uniform only with regard to these privileged systems, and the body is only controlled by the influence of its **inertia,** i.e. of its tendency to persist. On the other hand, an observer who refers such a motion to another system will always detect the presence of so-called **inertial forces,** a familiar example of which is **centrifugal force** and also the apparent force (Coriolis force) that gives rise to the easterly deflection of freely falling bodies on the rotating earth.[11] The effect of such inertial forces is also manifested in the jerk experienced when the brakes are suddenly applied to a railway train, or when we travel round a sharp curve, and also in the phenomenon of the **Foucault pendulum experiment,** or the **oblateness of the earth** that is referred to the earth's rotation.

Thus we feel tempted to deny the existence of absolute velocities, but to assume the existence of **absolute accelerations and rotations.** However, such ideas would in the long run signify a return to the prejudices surmounted by the principle of relativity, to the Newtonian conceptions of an absolute space and an absolute time. We can understand that theoretical physicists and in particular the creator of the theory of relativity could not permanently reconcile themselves to the idea of such a partial and inherently contradictory relativity. An ingeniously simple consideration brought Einstein on to the right path to the solution of this problem, although he had necessarily to surmount enormous difficulties before he reached his destination.

In order to understand Einstein's basal ideas, we shall imagine an observer **A**, and a body **M** which is moving solely under the influence of its inertia in a definite direction, which we shall call the **z**-direction. For the observer **A** the body will then of course be moving uniformly and in a straight line. Let us now consider a second observer **B**, who is moving

[11] At a place of geographical latitude 50°, the easterly deflection amounts to about 1½ cm. for a height of fall of 100 metres.

away from the first observer **A** in the **z**-direction with a **constant acceleration f.** To the observer **B** the motion of the body will appear to be **uniformly accelerated,** the acceleration being of magnitude **f** and in the opposite direction to the **z**-direction. We shall now consider that we have a number of bodies of different mass and of different chemical nature, bodies composed, say, one of lead, one of stone, one of cork, and so on; we imagine them to be moving in any direction and with arbitrary velocities, but only under the influence of their own inertia. As far as the observer **B** is concerned, all these bodies will exhibit the **same acceleration,** and this will be **independent of the mass and of the chemical nature,** and of magnitude **f** in the direction opposite to the **z**-direction. Provided he knows nothing of the observer **A** and of his own acceleration with respect to **A,** the observer **B** will obviously interpret the above result as **indicating** that he is situated in a **homogeneous gravitational field,** for it has been known to physicists since the time of Galilei to be characteristic of such a field, that all bodies situated in it experience the same acceleration, and that this is independent of all the properties of the bodies.

In the description of the physical processes that take place in the neighbourhood of the two observers, it is thus quite immaterial whether we take the point of view of the observer **A** who refers the processes to his co-ordinate system, or of the observer **B,** who makes use of his own co-ordinate system and in addition assumes the presence of a homogeneous gravitational field. Physical processes offer no possibility of deciding between the two points of view, and we must thus regard them as being equally justified.[12] This is generally known as **Einstein's principle of aequivalence,** and it immediately suggests that the problem of **gravitation** must be most closely linked with the problem of the general theory of relativity.

Having been guided in the right direction by the principle of æquivalence, we can now form some idea of the fundamental notions of the general theory of relativity. Like gravitational forces, so all **inertial forces,** being characteristic of the behaviour of a moving body at a particular point in space and at a definite time, are not a consequence of absolute accelerations, but are **determined by the space-time distribution of matter in the universe.** Conversely, however, this also opens up the possibility of another point of view, according to which the **law of gravitation** is nothing else than the **generalization of Galilei's law of inertia demanded by a truly general principle of relativity.**

How, then, is a generalization of this law possible? The law of inertia teaches us that every body moves in a straight path with constant velocity, so long as no outside forces act upon it. If we pass from three into four dimensions, the law of inertia thus indicates, according to what has already been said about world lines, that the **world line** which describes a **motion unattended by the application of force** must be a

[12] The example given by Einstein himself, of a physicist enclosed in a large chest somewhere in space, is particularly illuminating. The chest is suspended from a rope that leads over a pulley, and the other end of the rope is pulled downwards with constant acceleration. If he knows nothing of the world outside the chest, and nothing about the pulley and the rope, the physicist inside the chest will naturally detect a gravitational field, which, if we imagine the chest to be on the earth instead of in space, will be superimposed on the gravitational field of the earth.

straight line. Now is this statement capable of generalization? As a matter of fact, such generalization is possible on the basis of a **generalization of the geometrical foundations of physics.**

In order to understand this, we shall for the purposes of simplicity first consider two-dimensional geometry. There can be no doubt as to when a line drawn on a plane sheet of paper can be considered as a straight line. But a **two-dimensional geometry** is just as possible **on a spherical surface** as on a plane. In fact, we can have a two-dimensional geometry **on any curved surface whatsoever.** However, the question as to when a line drawn in such a surface can be considered a straight line is not so easily answered. Suppose we consider two points on a model of the earth, which lie on the same parallel of latitude but have widely differing values of longitude. (Since we are dealing with two-dimensional geometry, we are of course unable to leave the surface of the globe.) What are we to understand by the straight line connecting the two points on the surface of the globe? Is it the portion of the parallel of latitude between them? In order that this question may have a meaning, we must obviously first replace the vague conception of a "straight line" by a more general and well-defined one. What we require is the idea of a **geodetic line,** a conception already long known in geometry, and **defined** as the **shortest line joining two points on the surface.** On the basis of this definition we find, by means of a geometrical investigation, that the portion of the parallel of latitude between the two points in the above consideration by no means represents a geodetic line. In order to obtain such a line we must construct a great circle on the sphere to pass through these two points. For this purpose we can temporarily transfer the pole to one of the points, and choose from the manifold of meridians passing through this point the meridian which passes through the second point. It may at first appear paradoxical, but we can explain in this way the fact that an airman who wishes to fly from London to Australia would, if he were to choose the shortest course, take a route via Petrograd.

We can clearly conceive a curved surface, as a two-dimensional manifold "embedded" in one of three dimensions. In contrast to this, however, we cannot form any conception of a curved space, which in itself is of three dimensions. But the absence of the possibility of the human mind forming a conception of such a space offers no barrier to an abstract mathematical treatment. In the year 1827 **Gauss** developed a **general theory of surfaces,** and later, in 1854, the great mathematician **Riemann** showed how our usual **three-dimensional solid geometry** can also be regarded as a **special case** of a more general three-dimensional geometry, in which space is "curved" just as a surface may be curved in two-dimensional geometry. Moreover, and Riemann already recognized this, what holds for two-dimensional geometry is equally true for every geometry, **no matter what the number of dimensions** may be. The same must therefore also apply in the four-dimensional manifold known as the **Minkowski world,** which links together space and time. When we make use of a general instead of a special geometry as the basis of its mathematical treatment, we must regard the Minkowski world also as being **curved,** and of course the degree of curvature may vary continuously from place to place.

We owe to Einstein the ingenious idea that the curvature of the Minkowski world is closely related to the space-time distribution of matter. Einstein's **generalized law of inertia,** which at the same time represents the new **law of gravitation,** states nothing more nor less than that every body always moves in such a way that its **world line** is a **geodetic line in the curved Minkowski world,** the **curvature** being **produced by matter.**[13]

This conception of gravitation leads directly to a very important conclusion. Let us suppose a body with any velocity to be situated at any position in space and time. We shall consider the mass of this body, however, to be only a "test mass," i.e. it is so small that it is not capable of influencing the field of matter nor the curvature of the Minkowski world. From what has already been said, it follows that the motion of the test mass is represented by a geodetic line passing through the world point under consideration, in the direction corresponding to that of the given velocity. Now let us imagine a second test body to be situated at the same place, at the same time, and to possess the same velocity as the previous one ; but in this case we shall suppose that its **mass is only half as large** as the mass of the first body. Here the motion will be represented by **exactly the same geodetic line** as in the first example. As previously shown, however, the acceleration is completely determined by the world line, for a given system of reference. Hence if we regard the law of gravitation as a generalization of the law of inertia,

[13] In this connection it is assumed that no external forces act on the body. Electromagnetic forces would come under this category, but, on the new conceptions, so-called gravitational forces would no longer be included.

we are led to the inevitable conclusion that the **acceleration**, which **we interpret as an effect of gravitation**, must be **independent of the mass.**

We can express this result in another way. It is usual to define the **inertial mass** of a body as the constant ratio that exists between an arbitrary force acting on the body and the acceleration resulting from this force. (We shall not take account of the relativity correction here.) We thus arrive at a unit of mass, the magnitude of which depends on the units used for the force and the acceleration. On the other hand, we can also determine the mass of bodies by measuring the force of attraction they experience in the field of gravitation, in terms of the force exerted on our unit of mass in the same field. The number representing this ratio then determines the **gravitational mass** of the body. If, now, the acceleration in the gravitational field is actually independent of the inertial mass, then according to Newton's second law of motion the force of gravitation must be exactly proportional to the inertial mass. Hence an inertial mass of twice the previous magnitude must correspond to twice as large a value of the gravitational force, and thus to twice the **gravitational** mass, and so on. In general, then, **gravitational and inertial mass** must be **identical.**

This conclusion already found definite expression in Galilei's observation that all bodies, whatever their mass, fall at the same rate in empty space. In classical mechanics it was taken for granted that the inertial mass of a body is proportional to the body's weight. The correctness of this assumption was confirmed in detail by accurate measurements which **Eötvös** carried out in 1890. These measurements were based on the fact that the weight of a body represents the resultant of two forces, one of which is the attraction of the earth, and the other the centrifugal force produced by the rotation of the earth. Whereas the former of these is determined by the gravitational mass of the body, the latter depends on its inertial mass. The experiments of Eötvös were carried out with extraordinary precision, and they showed that there was no detectable difference in the direction of the resulting force, even when the masses differed widely in magnitude and in chemical constitution. The law of the identity of gravitational and inertial mass appears to be completely established by these measurements. But whereas this law was regarded as self-evident prior to Einstein's theory, and physicists gave it no further thought, it was Einstein's work that revealed the profound reason for this remarkable law.

But the new law of gravitation is of course still incomplete, so long as we do not know the relation which expresses the way in which the **Minkowski world** is **curved by matter.** In order to arrive at a solution of this question, we shall imagine a **dual system of numerous curves,** which we shall distinguish as **x**- and **y**-curves. In this system of curves, no **x**-curve shall cut another **x**-curve, and none of the **y**-curves intersect any other **y**-curves (Fig. 6). We can then imagine each of the continued series of **x**-curves denoted by successive whole numbers, and similarly for the **y**-curves. But we can also imagine nine intermediate curves drawn between, say, the curves $x = 6$, and $x = 7$, which in their turn shall not intersect each other, and we can denote these curves by the numbers 6·1, 6·2, etc., to 6·9. Furthermore, we can insert between the curves $x = 6·1$ and $x = 6·2$ nine fresh curves, and give these the numbers 6·11, 6·12, etc., to 6·19. In this way we can finally denote every point on the surface by two numbers corresponding to its respective **x**- and **y**-values, and regard these in the broader sense of the word as the co-ordinates of the point in question. These values are called the **Gaussian co-ordinates** of the point. The **Cartesian** co-ordinate systems used in plane geometry represent, of course, only **special cases** of Gaussian co-ordinate systems.

After we have thus fixed a Gaussian co-ordinate

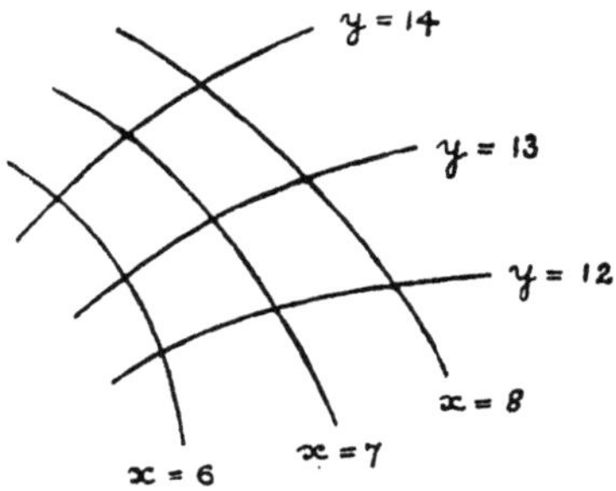

FIG. 6.

system, let us now consider three neighbouring points in the surface which make up a **triangle**. In the particular case of plane geometry, the co-ordinates of the three points would enable us to determine completely the lengths of the sides of the triangle, its angles and its surface area. In the geometry of a quite arbitrarily curved surface, however, this is not the case. The statement of the co-ordinates does **not** then suffice for the determination of the lengths of the sides, the angles, and the surface area of the triangle. **In addition to this**, as was recognized by Gauss, we must also be given a magnitude known as the

metrical fundamental tensor. With reference to a definite co-ordinate system in the surface this tensor is given by three **components.** The triplicity is to be explained by the fact that the tensor has one component for each co-ordinate axis, and one as it were for the co-ordinate surface.

For the particular place at which we considered the small triangle to be constructed, it is absolutely necessary that the above-mentioned tensor be given, if we wish to calculate the sides, the angles, and the area of the triangle from the co-ordinates of the three points. The **measure relations**[14] are determined by this tensor; for this reason it enters necessarily into all the formulæ of a general theory of surfaces. Conversely, we can also determine the fundamental tensor by measuring up the triangle, and for this purpose we require to know nothing as to how the surface is "embedded in space" at the place in question. Only when the fundamental tensor is given for every position on the surface is a **general geometry of surfaces** possible. Only then has such a geometry any meaning, and of course the fundamental tensor may in general **vary in a continuous manner** from place to place on the surface.

In general the number of components of the fundamental tensor is equal to the sum of the number of co-ordinate axes and co-ordinate surfaces. In three-dimensional geometry the fundamental tensor thus has three plus three or six, and in four-dimensional geometry four plus six or ten components. This follows because there are six co-ordinate planes in four-dimensional geometry; one is formed by the first and second co-ordinate axes, one by the first

[14] For this reason it is called the "metrical" fundamental tensor.

and third, one by the first and fourth, one by the second and third, one by the second and fourth, and finally one by the third and fourth.

If physics frees itself from the arbitrary **prejudice** that geometry in the Minkowski world must be as it were plane, or free from curvature, then **of geometrical necessity** it must ascribe to every position in the Minkowski world a definite value of the fundamental tensor of ten components, and this tensor can vary only continuously from position to position. The Minkowski world thus appears as the field of its own fundamental tensor. On the other hand, the space-time distribution of matter, or in other words the space distribution of matter and its condition as regards velocity is likewise representable by a so-called tensor of ten components, which also has its definite value at every place in the Minkowski world.

By means of a few plausible assumptions, Einstein arrived at a definite relation that links up this so-called tensor of matter with the fundamental tensor, and with certain magnitudes which can be derived from the fundamental tensor by purely mathematical operations. This relation is expressed by **Einstein's** so-called **field equations**; according to these, Einstein's law of gravitation states that the motion of a body is represented by a geodetic line in the Minkowski world, and the curvature of this line, which is determined by matter, must further satisfy the field equations.

Since both the geodetic line and the world curvature are completely independent of the Gaussian coordinate system used, it follows that Einstein's law of gravitation, or in other words the generalized law of inertia also holds **for any Gaussian co-ordinate system whatsoever.** On the other hand, every change of the system of reference, which in space appears as a transition between two three-dimensional co-ordinate systems **moving quite arbitrarily with respect to each other**, is always represented in four-dimensional "space" by a transition from one to some other Gaussian co-ordinate system.

The generalized law of inertia, when in the form of Einstein's law of gravitation, must therefore hold in exactly the same way for two different spatial co-ordinate systems, whatever may happen to be the motion of these with respect to each other. For the transition from one system of reference to another, only the **components** of the tensor magnitudes occurring in the law vary, and it is just the alterations thus connected with the transition that were interpreted by classical mechanics as **inertial forces**, which are different according to the choice of the system of reference. The postulate of relativity thus appears fulfilled in the most complete manner by the new theory. For this reason Einstein's theory of gravitation is also called the **general theory of relativity**, in contrast to the earlier Einstein theory known as the **special theory of relativity** (because it appears in the new theory as a special case).

Newton's familiar **law of attraction**, i.e. the law according to which the force of attraction between two bodies is directly proportional to the product of their masses, and inversely proportional to the square of their distance apart, follows **to a first approximation** from Einstein's law of gravitation. However, Newton's law represents only a **special case of Einstein's law of gravitation**, which must, of course, be much **more general** than Newton's law.

Now there are certain phenomena the occurrence

of which is a necessary consequence of Einstein's law, whereas they are **not** to be expected on the basis of Newton's law. As a result of this, we are presented with the possibility of putting the **general theory of relativity to the test of experience.** In contrast to Newton's law, in the first place, the law of Einstein leads to the important conclusion that the **ellipses** which the individual planets describe round the sun must undergo a continuous, but very slow **rotation** (in their own planes), the magnitude of this rotation being greater, the nearer the planet is to the sun. In point of fact, the astronomer **Leverrier** had already discovered, in the middle of the nineteenth century, that the orbit of **Mercury**, the planet nearest the sun, turns through an angle of 43 seconds of arc in a century; i.e. the rotation actually observed is greater by that amount than it ought to be in virtue of the attraction of all the heavenly bodies concerned. This anomaly had been a complete puzzle to astronomers, and it had even been concluded from this observation that an unknown planet must exist between the orbits of Mercury and the sun, and that the supposed disturbances of the orbit of Mercury were caused in this way. Now from Einstein's theory, not only does the fact of the rotation of the orbit of Mercury follow, but the value obtained by calculation is actually found to be exactly the amount observed, or 43 seconds of arc per century. The derivation of Newton's law of gravitation as a special case was a great accomplishment of Einstein's theory, but **this** result was a still greater achievement. Einstein's theory received wonderful confirmation in its application to the orbit of Mercury, for the agreement was complete in both its qualitative and its quantitative aspects, and this phenomenon was the only one previously known to astronomers that stood in contradiction to Newton's theory.

From Einstein's theory it also follows that **spectral lines** originating in **stars of very large mass** ought to show a **displacement towards the red end of the spectrum.** The magnitude of this effect lies at the limits of accuracy of present-day measurements. For this reason a definite decision as to this effect has not yet been possible, although some measurements of **Grebe** and **Bachem** seem to provide some evidence for the probable existence of the effect.[15]

A third deduction to which Einstein's law of gravitation leads has been found to be exactly as predicted by theory. This prediction, although it at first sight appeared a curious one, was confirmed in the year 1919 by means of astronomical observations, the results of which were awaited with great interest by followers and opponents of the theory of relativity alike. It follows from Einstein's theory of gravitation that **rays of light** must suffer **curvature** when they pass close to large masses. A ray of light from a star that just grazes the surface of the sun in its journey to the earth ought, according to Einstein's theory, to experience a total **bending** of 1·7 seconds of arc. If it passed the sun at a distance equal to twice the sun's radius from the centre of the sun, the curvature would be half as large, and so on. In consequence of the curvature of the rays, the star emitting them would not appear to a terrestrial observer to be in the position actually occupied by it. Thus the stars which appear to be situated in the sky in the neighbourhood of the sun must **seem** slightly **displaced in the direction away from the sun,**

[15] See the lecture of Grebe deliveredat the "Naturforschertag" at Nauheim, 1920; "Physikalische Zeitschrift," **21**, p. 662 (1920).

the amount of this displacement being greater, the nearer the star is to the sun (Fig. 7).

Now the stars which are situated in the neighbourhood of the sun are in general invisible; but during **a total eclipse of the sun** they can be observed and photographed. If, then, the "Einstein effect" actually exists, it ought to reveal itself during such a solar eclipse. Since a total eclipse of the sun was expected on the 29th May, 1919, in equatorial regions, two English expeditions were equipped and made observations at Sobral in Brazil and on Principe Island off the west coast of Africa. During the eclipse they photographed the sun and the surrounding fixed stars, and when the plates were accurately measured, they showed the effect predicted by Einstein. The agreement was complete, not only qualitatively, but quantitatively to within a limit of error of about 10 per cent.[16] This was in truth a wonderful triumph for this boldest of all theories, the general theory of relativity.[17] (See p. 154.)

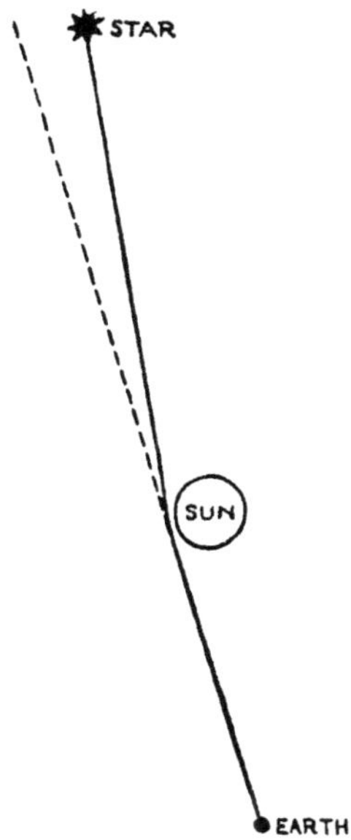

FIG. 7.

[16] See Appendix III in Einstein's book, "Relativity: The Special and the General Theory." Methuen, London, 1920.

[17] In 1911, before the publication of his work on the general theory of relativity, Einstein had already predicted one-half of the effect actually observed in the year 1919. In 1914, a German expedition was to have looked for the predicted effect, during the total eclipse of the sun which took place in August of that year. Their intentions were frustrated, however, by the outbreak of war.

But the new theory was to reveal its immense importance in yet another direction. On the basis of his theory of gravitation, Einstein in 1917 succeeded in dealing satisfactorily with one of the loftiest problems of natural science, the question as to the **finiteness and the magnitude of the universe.**

Since the earliest times, philosophers and scientists alike have been occupied with the question as to whether our universe is finite or infinite. Although the majority of people were of the opinion that there must be an infinitely large number of stars, physicists and astronomers recognized decades ago that the **assumption of an infinite universe** is **incompatible with Newton's law of gravitation.**

Let us suppose space to be infinite and uniformly filled with stars. In such a case, according to classical mechanics, a quite indeterminable force would act on every heavenly body **from all directions,** and a stable arrangement such as we have in our solar system would then be quite impossible. If, on the other hand, we imagine the universe to be spatially infinite, but to have a finite number of stars, it would again remain unintelligible why the stars do not lose themselves in the infinite universe. Thus the universe cannot be infinite, and we can nevertheless hardly imagine it to be limited in any way. These difficulties can be surmounted if we make the assumption that the **universe** may be **unbounded,** but that it has a **finite volume.**

Such an assumption only appears to be paradoxical, for a **circle** also has no beginning and no end, and yet it has an exactly determinable length. In spite of its unlimitedness, a circle is finite, because it is **curved.** A traveller on the surface of the earth will nowhere find a wall or an abyss, which mark the end of the earth, and yet the earth's surface has a perfectly definite and finite area. Since a great circle of the earth is a continuous line,[18] it follows that a **geodetic line** drawn on the earth's surface has quite a definite and expressible length, which represents the **circumference** of the earth and has the value of about 40,000 km. The three-dimensional analogy to the one-dimensional circular line and the two-dimensional spherical surface would be a so-called **spherical space**, which, however, proves itself to be quite inaccessible to human perception, although such a space offers no difficulties to an abstract mathematical treatment such as the previously mentioned geometry of Riemann. But we must beware of all attempts to conceive of spherical space; they are fruitless, and attempts of this kind only lead to the fundamental error of confusing spherical space with a sphere.

In consequence of the principle of relativity, the idea of space has lost its independence. However, on the basis of the important empirical fact that the **relative velocities of stars** are **small** as **compared with the velocity of light**, we can, to a certain approximation, split up the Minkowski world into space and time. We can thus speak approximately of a **space**, and this Einstein regards as **spherical** in the sense already mentioned, in that it is boundless, but has a finite volume and finite girth.

Incidentally, the idea that our universe, in spite of its boundlessness, might be finite in consequence of an existing curvature, was probably expressed

[18] The geodetic line can also be defined as the curve described when we advance from one point to an immediately contiguous one, and continue doing this without ever changing the direction of advance.

before Einstein's time. But with Einstein the conception of a curvature of the universe is not an arbitrary one; it results of necessity from his new theory of gravitation, in that the curvature appears to be brought about according to a definite law by the matter distributed in the universe. In this way we obtain a perfectly definite connection between the world curvature and the world mass, and, on the other hand, the mutual attraction between very small spheres of lead in a terrestrial laboratory must again depend on the girth and mass of the whole universe, so that these two cosmological constants must manifest themselves in the intensity of the attraction. Thus **Einstein's cosmological theory of gravitation** led to definite values for the size and the mass of the universe.[19]

From this theory we obtain two simple equations that link together in all five magnitudes, two of which are well-known, two unknown, and one of which the order of magnitude can be estimated. By solving the two equations for the two unknowns, the order of magnitude of these can also be estimated; but these two unknowns are none other than the size and the mass of the universe. The two well-known magnitudes are the velocity of light and the **gravitational constant,** i.e. the universal constant that expresses the force with which two bodies each of mass 1 grm. attract each other at a distance of 1 cm.[20] The magnitude that can only be estimated is the

[19] In order to arrive at the cosmological applications of the theory of gravitation, however, Einstein had still to add a supplementary term in his "field equations," but this is not of importance for the derivation of Newton's law as a special case, nor of the three previously discussed effects of the general theory of relativity.

[20] Cf. note 8 of the third lecture.

average density of matter in the universe, or the quotient of the total mass of the universe and its total volume.

The average density of matter in the universe is found by making an estimate of the space in which the observable fixed stars are contained, and further, by estimating the number of these fixed stars. (In this connection the average mass of a star is taken as being about equal to that of the sun.) Finally, account is also taken of the fact that the stars are distributed in star clusters, from amongst which our milky way system only represents one, whereas the others appear to us as nebulæ. In this way, the astronomer **De Sitter**[21] reached the conclusion that the average density of matter in the universe is about 100 quadrillion times smaller than the density of water. This is not to be wondered at, when we consider how small the linear dimensions of the heavenly bodies are, as compared with the distances in the stellar universe. Although the sun contains almost the whole mass of our solar system, the distance between it and the nearest fixed star (Alpha Centauri) is nevertheless about 60 million times larger than the radius of the sun.

On the basis of this estimate of De Sitter, the **circumference of the universe** comes out to be about **100 million light-years.** (A light-year is the distance that light would travel in the course of one year.) De Sitter's estimate certainly sounds tremendous; yet according to the estimates of astronomers, on the other hand, this enormous length is only about ten times as large as the distance from

[21] W. De Sitter, "On Einstein's Theory of Gravitation, and its Astronomical Consequences." Monthly Notices of the Royal Astronomical Society, London (1917).

the earth to the farthest spiral nebula visible in the telescope.

On De Sitter's estimate the **mass of the universe** is found to be about **a quintillion quadrillions** (10^{54}) **of grams.** It bears about the same relation to the mass of the earth, as the mass of the earth bears to that of a stone, so that we can now extend what we said in a previous lecture in reference to the mass of atoms, and form a triple proportionality between the masses of an atom, a stone, the earth, and the universe, in the following way: **an atom is to a stone as a stone to the earth, and as the earth to the universe.** The mass of the universe works out at about a 100 trillion times the mass of the sun.

Another interesting point arises from our previous considerations, and this is the question as to the **total number of atoms in the universe,** or in more precise terms, the total number of **electrons,** since they are the fundamental bricks of matter. Now there are about a quadrillion electrons in a gram, so that the number of electrons in the universe would be of the order of magnitude of a **quintillion octillions** (10^{78}), i.e. ten raised to the 78th power. Of course such a number far exceeds everything that humanity can conceive of, but at the same time a simple consideration shows that in thought this number can quickly be arrived at by means of an exercise in a calculus of combinations, as shown below.

It is well-known that bacteria are peculiar in that they propagate by a process of division of the individuals. From one bacterium in the course of an hour there result two bacteria by this splitting-up process, and after an interval of two hours there are four, and so on. Let us suppose we have a single bacterium in a glass of water, and that it has a thickness of about a thousandth of a millimetre and twice this length. It would thus be approximately a billionth part of a gram in weight, and made up from roughly a billion electrons. We shall further assume that in some way or other sufficient nourishment can be supplied to ensure that reproduction is not adversely affected by lack of food. In such circumstances there would be present some 16 millions of bacteria after one day, i.e. twenty-four hours; at the end of the second day there would be 300 billions, and after three days about 5000 trillions, which would already correspond to a weight of thousands of tons. In the course of the sixth day the mass of the bacteria produced would exceed the mass of the earth; in the course of the seventh day the mass of the sun; in the course of the tenth day the weight of all the bacteria would attain to the total weight of the universe; and finally in the course of the eleventh day the number of all the bacteria that would have developed from the original one would be as large as the total number of electrons in the universe.

Thus, even though the **universe** is immensely large, it is nevertheless by no means infinite; on the contrary, **its magnitude is comprehendable by the human mind.** Mankind certainly appears in the highest degree subordinate in the vast step-ladder that leads us from the atom to the universe; but how immense his mind appears, when we consider that he has succeeded in advancing in theoretical knowledge to the smallness of atoms on the one hand, and to the immensity of the universe on the other, and all this in spite of the limitations of the human senses! A unified system of explanation of nature leads in the modern system of physics from the ultimate primordial particle to the universe in its entirety.

Positive and negative electrical charges now appear as the bricks of the universe. They are arranged in numerous ways into systems that we call atoms. The kind of arrangement determines the **chemical** nature of the atoms, the internal structure of which is regulated by quantum relations. The appearance of a **mechanical** mass follows from electrical properties. In consequence of their charge and their motion, the electrons call forth an electromagnetic field, which is periodic in space and time. Space is therefore filled with electromagnetic waves of all possible frequencies. Only a very small part of the spectrum is revealed to the most perfect of the human sense organs, i.e. to the eye, as **light.** Totality phenomena, in which a large number of material individuals take part, are responsible for the phenomena of **heat.** Space and time, however, in which all processes appear to operate, are linked together to form a four-dimensional manifold, the geometry of which is manifested in the phenomenon of **gravitation.**

Thus the new physics reveals to us a picture of nature of great simplicity. In point of fact, it is not nature that is complicated, but only the path that led to the true knowledge of it. This path was complicated because it started off from the narrow limitations of the human senses, and because theoretical physics only gradually succeeded in liberating it from human points of view.

Note added, June, 1923.—Subsequent observations made during the solar eclipse of September, 1922, by three expeditions, have substantiated the 1919 results, and the truth of Einstein's prediction of the bending of light in gravitational fields can now be regarded as completely established.

HISTORICAL SUMMARY

1820. Oersted discovered the phenomenon of electromagnetism.
1831. Faraday discovered induced currents.
1833. Faraday discovered the fundamental law of electro-chemistry.
1842. Mayer discovered the constancy of the transformation ratio between heat and mechanical work.
1845. Faraday discovered that magnetism is a universal property of all substances.
1856. Weber found an electrical constant equal to the velocity of light.
1857. Clausius founded the newer kinetic theory of gases.
1859. Kirchhoff formulated the fundamental principle of the theory of radiation.
1859. Kirchhoff and Bunsen founded spectroscopy.
1859. Plücker discovered cathode rays.
1865. Loschmidt first estimated the size of molecules.
1866. Boltzmann explained irreversibility from the standpoint of molecular statistics.
1873. Maxwell founded the electromagnetic theory of light.
1876. Rowland proved experimentally the existence of convection currents.
1879. Stefan discovered the law of radiation which bears his name.
1881. J. J. Thomson introduced the conception of electromagnetic mass.
1885. Balmer discovered the law of the spectrum of hydrogen.
1888. Hertz confirmed Maxwell's theory by means of his experiments.
1893. Wien discovered the displacement law.
1895. Lorentz founded the electron theory.
1895. Röntgen discovered the rays which bear his name.
1896. Becquerel discovered the phenomenon of radioactivity.
1896. Zeeman discovered the magneto-optical effect which bears his name.
1898. M. and Mme. Curie discovered radium.
1900. Planck founded the quantum theory.

1900. Planck introduced the general law of radiation.
1900. Planck determined the elementary quantum of action, and first calculated exactly the mass of the atom of hydrogen and the elementary quantum of electricity.
1900. Rutherford discovered thorium emanation.
1902. Rutherford and Soddy enunciated the theory of atomic disintegration.
1905. Einstein formulated the principle of relativity.
1905. Einstein discovered the law of the inertia of energy.
1905. Einstein founded the theory of light-quanta.
1906. Nernst introduced a new theorem of heat.
1907. Einstein founded the quantum theory of heat of solid bodies.
1908. Minkowski introduced the conception of a four-dimensional world, linking together space and time.
1912. Laue discovered the interference of Röntgen rays.
1913. W. H. and W. L. Bragg discovered the structure of crystals.
1913. Moseley founded Röntgen-spectroscopy.
1913. Bohr founded the quantum theory of spectra and of atomic structure.
1915. Einstein founded the general theory of relativity.
1915. Einstein and de Haas proved experimentally the existence of molecular currents.
1915. Sommerfeld explained the fine structure of spectral lines.
1919. Rutherford discovered the formation of hydrogen from nitrogen.
1919. Aston discovered isotopy in non-radioactive elements.

NAME INDEX

(In the case of deceased scientists the professorships mentioned in this index are those they occupied last In the case of living scientists the data refer to the posts they occupied in 1922. When not otherwise stated, the data refer to posts in physics. In this index no account has been taken of the footnotes in the text.)

SUBJECT INDEX

PRINTED IN GREAT BRITAIN AT THE UNIVERSITY PRESS, ABERDEEN

Bei Fragen zur Produktsicherheit wenden Sie sich bitte an:
If you have any questions regarding product safety,
please contact:

Walter de Gruyter GmbH
Genthiner Straße 13
10785 Berlin
productsafety@degruyterbrill.com